21世纪软件工程专业教材

SOFTWARE TEST PRACTICE TUTORIAL

软件测试实践教程

主 编 王智钢 王蓁蓁

副主编 曾 岳 张海涛 李 莉

清华大学出版社

北 京

内 容 简 介

本书全面介绍了软件测试的产生与发展,软件缺陷与软件事故,软件测试的意义、原则和挑战,软件测试模型、过程和组织管理等基础知识,并着重介绍了动态白盒测试设计与动态白盒测试实践,黑盒测试设计与黑盒测试实践,结合全国大学生软件测试大赛的题目等讲解 Web 测试方法。

本书分为四部分。第一部分为第 1 章绪论,简要介绍软件测试的基础知识;第二部分为第 2~4 章,重点介绍白盒测试的方法技术和实践应用;第三部分为第 5~6 章,重点介绍黑盒测试方法的技术和实践应用;第四部分为第 7 章 Web 测试,结合全国大学生软件测试大赛介绍 Web 测试方法。全书以大量源程序代码和测试代码作为示例,每章均附有丰富新颖的习题。

本书的读者对象为软件工程、计算机科学与技术等相关专业国际教育类学生。

图书在版编目(CIP)数据

软件测试实践教程 = Software Test Practice Tutorial:英文/王智钢,王蓁蓁主编. --北京:清华大学出版社, 2025.8. -- (21 世纪软件工程专业教材). -- ISBN 978-7-302-69880-7

Ⅰ. TP311.5

中国国家版本馆 CIP 数据核字第 2025LE3586 号

责任编辑:张 玥
封面设计:常雪影
责任校对:王勤勤
责任印制:沈 露

出版发行:清华大学出版社
　　　　网　　　　址:https://www.tup.com.cn,https://www.wqxuetang.com
　　　　地　　　　址:北京清华大学学研大厦 A 座　　　邮　　编:100084
　　　　社　总　机:010-83470000　　　　　　　　　邮　购:010-62786544
　　　　投稿与读者服务:010-62776969,c-service@tup.tsinghua.edu.cn
　　　　质　量　反　馈:010-62772015,zhiliang@tup.tsinghua.edu.cn
　　　　课　件　下　载:https://www.tup.com.cn,010-83470236
印　装　者:三河市君旺印务有限公司
经　　销:全国新华书店
开　　本:185mm×230mm　　　印　张:13　　　字　　数:277 千字
版　　次:2025 年 8 月第 1 版　　　　　　印　　次:2025 年 8 月第 1 次印刷
定　　价:49.80 元

产品编号:100980-01

前　　言

　　基于先发国家科学研究、信息产业发展和技术实践的英文版软件测试教材不少，国内把英文版译成中文版或者编写中文版软件测试教材的也不少。但站在我国的视角，以中英文双语为载体，涵盖经典软件测试教学内容，并反映我国在软件与信息技术服务产业发展、信息技术应用创新中的软件测试技术实践和技术创新等内容的教材还不多。随着教育国际化的蓬勃发展，"软件测试"作为计算机类专业的一门重要课程，需要双语教材来支撑课程教学和人才培养。这本双语教材在帮助国内学习者对接学习国外软件测试课程的同时，也能够帮助国外学习者对接学习国内软件测试课程。

　　本书以 IT 企业对研发人员的软件测试技术能力要求为导向，以软件测试工程能力培养为目标，梳理知识单元和能力要素，组成知识体系和能力体系；理论和实践有机结合，以大量源程序代码和测试代码作为示例来讲解，便于读者学习和掌握，促进实践能力产出。本书可以作为软件工程、计算机科学与技术等相关专业国际教育类学生软件测试课程的教材或参考书。

　　全书共 7 章，章节安排围绕黑盒测试和白盒测试两大主线，既有理论讲解，也有应用案例。第 1 章为绪论，简要介绍软件测试的基础知识；第 2 章为静态白盒测试，介绍代码检查、编码规则和编程规范、静态质量度量等，并给出静态白盒测试实例；第 3 章为动态白盒测试设计，介绍逻辑覆盖、基本路径覆盖、循环测试、变异测试、符号执行以及程序插桩和调试；第 4 章为动态白盒测试实践，介绍 JUnit 单元测试，给出逻辑覆盖实例、面向对象多态测试实例；第 5 章为黑盒测试设计，介绍等价类划分、边界值、错误推测、判定表驱动等；第 6 章为黑盒测试实践，介绍自动化黑盒测试的基本原理和相关技术；第 7 章为 Web 测试，介绍 Web 自动化测试，结合全国大学生软件测试大赛的题目给出实践案例。

　　本书具有以下特点。

　　（1）面向产业能力需求，注重实践能力产出。将软件与信息技术产业对研发人员的软件测试技术要求作为能力目标，分析知识要素和能力要素，组织理论知识体系和实践内容体系，结合软件测试项目应用和学科竞赛，促进软件测试分析、测试设计、测试开发和测试执行实践能力产出。

　　（2）理论学习和实践应用相结合。以大量源程序代码和测试代码作为示例讲解软件测试知识，并提供完整应用案例，帮助读者学以致用。

　　（3）提供丰富新颖的习题，促进巩固提高和融会贯通，加强对能力产出水平的度量和考核。

（4）提供配套的教学大纲、教学课件、程序源码、习题答案、案例素材等，读者可在清华大学出版社官方网站下载。

本书由王智钢、王蓁蓁、曾岳、张海涛、李莉共同编写。其中，王智钢编写了第 1、3、4 和 5 章并统稿，王蓁蓁编写了第 2 章，曾岳编写了第 6 章，张海涛、李莉编写了第 7 章和各章习题。在编写过程中，本书用到全国大学生软件测试大赛的题目资源，也参考了国内外教材和资料，对相关作者表示由衷的感谢。本书在出版过程中，还得到了清华大学出版社的大力支持，在此表示诚挚的感谢。

由于作者水平有限，书中难免有不妥和疏漏之处，恳请各位专家、同仁和读者不吝赐教。

王智钢

2025年5月于南京

目　　录

Chapter 1   Introduction

1.1   The emergence and development of software test

1.1.1   The emergence of software test

FORTRAN is the first high-level language to be officially popularized in the world. It was proposed in 1954 and officially used on October 15, 1956. The following is a FORTRAN language program code example "Solving quadratic equations".

```
     PROGRAM QUADRATIC
50   READ(*,10)A,B,C
10   FORMAT (3F3.1)
     IF (A. EQ.0.0) THEN
      WRITE (*,*) ' The quadratic coefficient cannot be 0'
      STOP
     ENDIF
     D=B*B-4.0*A*C
     IF (D. LT.0.0) THEN
      WRITE (*,*) ' This quadratic equation has no real roots'
      STOP
     ENDIF
     X1=(B+SQRT(D))/(2.0*A)
     X2=(-B+SQRT(D))/(2.0*A)
     WRITE(*,20) X1,X2
20   FORMAT(1X, F6.2,10X,F6.2)
     GO TO 50
     END PROGRAM QUADRATIC
```

In early days, most software was developed by the person or organization that used it, and there was no systematic software engineering method to follow.

The software development process was very casual, and the software developed often had a strong personal color. At the same time, the software was small in scale and low in complexity. Besides the source code, there was no software manual or other documents. In any case, the software had to go through a trial run and be modified possible errors before it can be put into actual use, otherwise it might lead to serious consequences.

Software test comes along with the generation of software. In early days, software test was basically equivalent to "debugging" and was usually done by the developers themselves. For software test in early days, the overall investment was very little, the test work started late, generally until the program code was written and the product was basically completed.

软件测试伴随着软件的产生而产生

It was not until 1957 that software test began to distinguish itself from debugging, becoming an activity dedicated exclusively to finding software defects. Initially, the purpose of software test was perceived as "to convince oneself that the product can works", so software test was usually done after the program code had been written. At that time, there was also a lack of effective test methods, relying mainly on "error guess" to find defects in software.

直到1957年,软件测试才开始与调试区别开来

1.1.2 The first class of software test method

软件测试的第一类方法

In 1972, Dr. Bill Hazel, a pioneer in the field of software test, organized the first formal conference on software test at the University of North Carolina. In 1973, he initially defined software test as the process of establishing confidence that the program will function as intended. In 1983, he revised the definition to "Evaluate the characteristics or capabilities of a program or system and determine whether it achieves the desired results. Software test is any act done for this purpose". In his definition, "function as intended" and "desired results" refer to what we now call user requirements or software specification. He also defined the quality of software as meeting the requirements. The fundamental concept underlying his thinking was that test was an effort to verify that the software was functioning as intended.

评价一个程序和系统的特性或能力,并确定它是否达到预期的结果

This is a positive thinking, for all the function points of the software system, verify its correctness one by one. The software test industry regards this approach as the first class of software test method. In 1975, John Good Enough and Susan Gerhart published the article titled "Principles of Test Data Selection", which identified software test as a research direction.

1.1.3   The second class of software test method

软件测试的第二
类方法

The first class of software test method had been questioned and challenged by many industry authorities, such as Glenford J. Myers. In 1979, Myers published his seminal book, *The Art of Software Testing*, which was considered the foremost and most significant monograph in the field of software test. He posited that test should not be centered on verifying that the software functions properly, but rather should begin with the assumption that the software contains faults, and then employ reverse thinking to uncover as many bugs as possible. Additionally, he argued from a standpoint of human psychology that if the purpose of test is to "verify the software works," it is highly detrimental for testers to discover faults in the software.

In 1979, Myers proposed a definition of software test: " test is the process of executing a program or system in order to find errors." This definition had gained recognition within the industry and is frequently cited. Myers also made three important points related to test as follows.

测试是为发现错
误而执行一个程
序或者系统的过
程

① Test is about proving that a program is faulty, not proving that it is error-free.

② A good test case is one that has the ability to uncover previously undiscovered errors.

③ A successful test is one that uncovers previously undiscovered errors.

This represents the second class of software test method, which tried to verify that the software is "not working", or there are bugs. Myers' notion that the purpose of test is falsification overturns the old notion that test is done to show that software is correct. Since then, the theory and method of software test have greatly evolved.

1.1.4   Understand two classes of software test method

However, Myers' concept of "the purpose of test is falsification" should not be interpreted too narrowly.

① If testers only focus on finding defects and seldom pay attention to the realization of system requirements, there will be some randomness and blindness in test activities.

② If some software companies adopt this view, measuring tester's performance solely based on the number of defects found would be unscientific. The value of test work does not solely lie in the number of defects found, and the amount of test efforts are not simply proportional to the number of defects found.

The goal of software test goes beyond just finding defects. It includes objectives such as objectively evaluating software quality and ensuring that a product meets certain quality standards. In general, the first class of software test method can be described as a process in which the software's functions are executed in an environment clearly defined by the software design. The execution results are compared to user requirements or design specifications, and if they match, the test passes, if not, it is considered a defect. The ultimate goal of this process is to run all functions of the software in all environments specified in the design and they all can pass.

The first class method is widely regarded as the mainstream and industry standard in the software industry. It is based on the requirements and design of the software, which helps to define the scope of test work and makes it easier to clarify its focus. This targeted approach is particularly important for large-scale software test, especially when time and human resources are limited. On the other hand, the second type of test method is not necessarily related to requirements and design. It emphasizes that testers should play a subjective initiative, with reverse thinking, constantly thinking about the misunderstanding of developers, bad habits, program code boundaries, invalid data input and various weaknesses of the system, trying to disrupt the system, destroy the system, and the goal is to find a variety of problems in the system. This method can often find more defects in the system.

1.1.5　From software test to software quality assurance

By the early 1980s, the software and information technology industry had entered a period of significant development. Software was becoming increasingly large-scale and complex, with a growing emphasis on quality. As a result, there was an increasing demand for more advanced test methods. During this time, fundamental theories and practical techniques of software test began to take shape, leading to the design of various processes and management methods for software development. The way of software development gradually transitions from disorder to structured development process, characterized by structured analysis and design, structured review, structured program design and structured test.

The concept of "quality" has also been integrated into the practice of test, leading to a change in the definition of software test. Test is no longer solely focused on identifying errors, but encompasses the evaluation of software quality as well. As a result, software test has become the primary method for ensuring software quality. In his book *The Complete Guide to Software Testing*, W. C. Hetzel emphasizes that "Test is any activity aim to assessing the characteristics of a program or system. Test is a measure of software quality".

测试是以评价一个程序或者系统属性为目标的任何一种活动。测试是对软件质量的度量

After this, software developers and testers began collaborating to discuss issues related to software engineering and test. In 1983, the Institute of Electrical and Electronics Engineers (IEEE) proposed a standard for software test in their Software Engineering Terminology, defining it as "the process of using manual or automated means to execute or measure a software system with the aim of verifying that it meets specified requirements or determining the difference between expected and actual results".

Software test is generally a method of checking after the fact. If the preliminary work of software development is not done well, it is difficult to rely entirely on test to ensure the quality of software products. In view of this, the combination of prior preventing, process monitoring, and result checking for software quality assurance has come into being. Software quality assurance aims to ensure that software products and services fully meet user requirements. Software quality assurance is a planned, organized activity that runs through the entire software process and includes the following.

结合事先预防，过程监督和事后检查的软件质量保证应运而生

① Identify software quality requirements and decompose them into quality elements that can be measured and controlled from top to bottom, which lays a foundation for qualitative analysis and quantitative measurement of software quality.

② Participate in software project planning.

③ Develop software quality assurance plan.

④ Monitor the software process.

⑤ Review software work products.

⑥ Conduct multi-level software test.

⑦ Review software project activities.

⑧ Generate reports on software quality assurance.

⑨ Deal with unqualified items and track problems.

Software quality assurance ensures that software products and processes adhere to established standards by defining quality criteria, developing quality assurance plans, implementing measures, conducting comprehensive supervision, performing stage inspections, providing quality reports, and tracking problem resolution. This enables the software process to be monitored, measured, and trusted by both the software project manager and the software user. The emergence and development process of software test is shown as Figure 1-1.

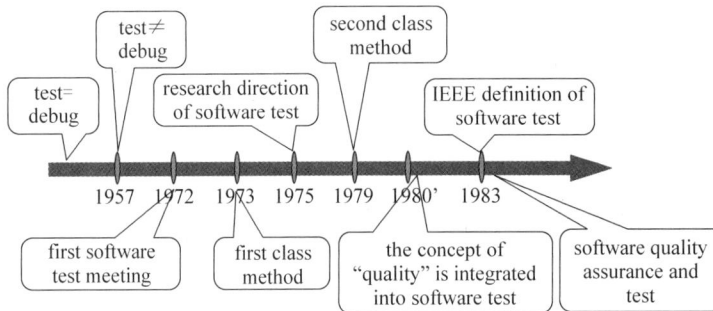

Figure 1-1   The emergence and development process of software test

In the process of the emergence and development of software test, the concept of software test is also constantly evolving and improving, which has roughly experienced four stages, as shown in Figure 1-2.

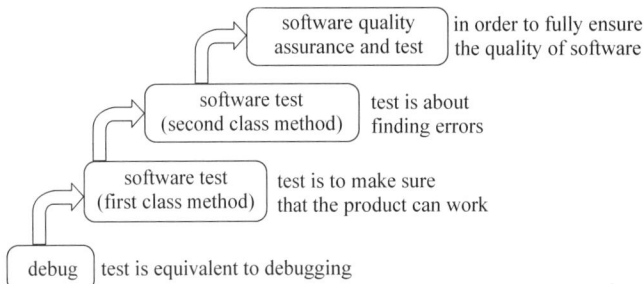

Figure 1-2   The development process of software test concept

1.1.6   The impact of software development characteristics on software test

The development of software has its own features, and these features have an influence on software test.

(1) In general, the number of software is increasing and the scale is getting bigger and bigger, which makes the task of software test heavier and heavier.

软件数量越来越多，规模越来越大，使得软件测试任务越来越重

With the development of the times, the number of software is increasing. Taking APP as an example, from the perspective of use, according to the "Operation of the Internet and Related Service Industries in 2020" published by the Ministry of Industry and Information Technology in February 2021, by the end of 2020, the number of APPs monitored in the domestic market was 3.45 million, of which the number of APPs in the local third-party APP store was 2.05 million. The number of APPs in the Apple Store (China) is 1.4 million. The total number of APPs developed should be greater than the number of APPs monitored in the market, because some might not enter the market and therefore not be monitored.

The size of software is increasing as well. For instance, the widely used Windows operating system consists of 45-60 million lines of code, the space shuttle has 40 million lines of code, and the space station has a staggering 1 billion lines of code. When other factors remain unchanged or change minimally, the number of software defects tends to be roughly proportional to the size of the software. For example, a software development team may observe that the defect rate per thousand lines of code in their developed products is 5. Therefore, when they develop similar software, they can estimate the approximate number of defects by multiplying the total number of lines of code by 5 per thousand.

(2) With the increasing complexity of software，the likelihood of defects also increases, making test more challenging.

软件复杂度越来越高，使得缺陷产生的概率增大，测试挑战度也越来越大

In 1962, Samuel, a pioneer in computer technology, developed a checkers program that defeated the U.S. state champion. In 1997, IBM's computer system Deep Blue defeated Kasparov, the world chess champion. In 2016, AlphaGo, developed by Google, defeated top professional player Lee Sedol. These three significant events (Figures 1-3) demonstrate our ability to develop increasingly complex software. Generally speaking, the number of defects in software is directly related to its complexity. As software becomes more complex, the probability of defects and test difficulty also increase.

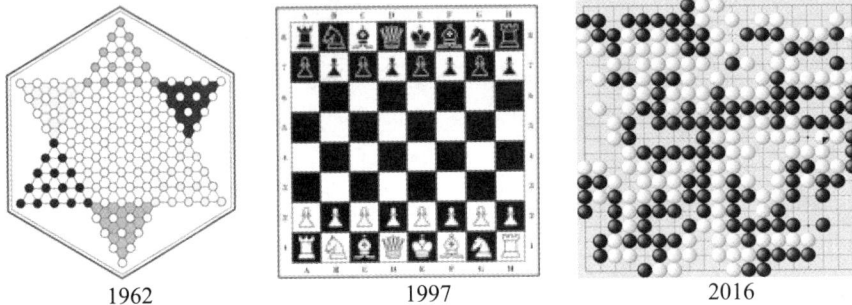

1962   1997   2016

Figure 1-3   Increasing software complexity

In May 2017, Google's autonomous driving team announced that their driver-less cars had undergone 8 years of test, accumulating a total mileage exceeding 4.83 million kilometers. This extensive test is equivalent to hundreds of years' worth of driving experience for an individual. Despite this progress, Google's driver-less cars still require ongoing test and cannot yet be deployed on a large scale for practical use. In March 2018, a tragedy struck when a self-driving test vehicles of Uber collided with and fatally injured a pedestrian during a road test in Tempe, Arizona, USA. This incident marked the first case ever of a self-driving vehicle causing harm to a pedestrian on public roads.

(3) With the rapid evolution of software application forms and hot spots, the requirements for software test are becoming increasingly diverse.

Taking payment applications as an example, there are various types such as bank card payments, online banking payments, Alipay, WeChat, etc., as shown in Figure 1-4. Furthermore, it can also be further evolved to include facial payments, sonic payments, etc., making the range of payment applications quite varied.

软件应用热点、应用形式快速演进，使得软件测试需求越来越多样化

Figure 1-4   Payment applications are varied

Nowadays, software can be categorized into standalone software, network software, cell phone APPs, embedded software, and other forms. The rapid evolution of hot spots and application forms in software applications has led to increasingly diverse needs in software test. Different types of software test require distinct knowledge bases, methods, and technical tools. Furthermore, new hot spots and forms of software may contain more defects due to technical immaturity and lack of experience accumulation. Therefore, it is essential to perform thorough test in order to ensure quality.

(4) As software applications continue to expand in both scope and depth, the field of software test is rapidly evolving to meet these new demands.

软件应用越来越广泛和深入，软件测试范围迅速扩大并深入，以满足这些新需求

With advancements in technology and increasing application requirements, the use of software has become more extensive and complex. It is no longer sufficient to simply understand software as a standalone system running on a computer. Many products now rely on software support or incorporate various software components, all of which require thorough test. The scope of software test has thus expanded beyond traditional boundaries to encompass all types of software products. For instance, as electric vehicles become more prevalent and autonomous driving technology advances, vehicles are increasingly resembling mobile computers. In order to ensure their safe operation, comprehensive and in-depth test is essential.

(5) The rapid evolution of software puts forward higher requirements for the response speed and execution efficiency of software test.

软件的快速迭代，对软件测试的响应速度、执行效率等提出了更高的要求

The competition of software market is becoming more and more fierce, and the requirements of users for software are becoming higher and higher. As a result, the pace of software maintenance and updates is getting faster and faster. For instance, new versions of mobile phone APPs are being released every 1-2 weeks. The rapid iteration of software has become a trend. This puts forward higher requirements for the response speed and execution efficiency of software test. Software test should be able to keep up with the rapid iteration rhythm of software versions and quickly complete test tasks for new versions.

(6) The application of software in important fields makes the requirement of software quality higher and higher, and the quality risk of software is increasing.

软件在重要领域的应用使得对软件质量的要求越来越高，软件的质量风险越来越大

Industries such as aerospace, weapons control, banking, and securities demand high levels of reliability and security in their software systems. Therefore, it is imperative to conduct thorough software test to ensure quality. In 2017, NIC Asia Bank in Nepal experienced fraudulent financial transfers through the SWIFT system, resulting in the theft of up to 460 million Nepalese rupees (approximately 47 million yuan). There have been several instances of SWIFT robberies in recent years, including a similar incident at the Bangladesh Central Bank in 2016, where a minimum of $81 million was lost.

1.2 Examples of software defects and accidents

软件缺陷和事故举例

1.2.1 Example of software defects

To get an intuition for what a software defect is, let's look at an example.

```
1  public class SalaryAverage
2   { public float getAverage ( String [] Salary )
3    { if (Salary==null || Salary.length==0)
4       { throw new NullPointerException(); }
5    float sum = 0.0F;
6    int j=Salary.length; for (int i=1; i
7    for (int i=1; i<j; i )
8      { sum = Salary[i]; }
9    return sum/j.
10  }    }
```

The code snippet above, which is used to calculate the average salary, has multiple flaws.

① Line 2 String [] scores should be int[] scores, because in line 8, Salary[i] requires arithmetic addition.

② Line 7 The initial value of the loop control variable i should be 0, because the Java array subscript is from 0 to the total number of array elements−1.

③ The code should be commented.

④ Lines 3 and 4, if the achievement array is empty or the length is 0, a specific message should be given.

1.2.2　Definition of software defects

软件缺陷的定义

Software defects are undesired or unacceptable deviations that exist in software (documents, data, programs). These defects may cause the software product that can't meet the needs of the user in some way. According to IEEE 729-1983, defects can be defined as errors, faults, and other problems that occur during the development or maintenance of the software product from an internal perspective. From an external perspective, defects refer to the failure or violation of a certain function that the system needs to achieve. The following situations are all considered software defects.

① The presence of errors in the software that are not supposed to occur according to the product specification.

② The failure of the software to fulfill a specified function outlined in the product specification.

③ The software functions beyond the scope specified in the product specification.

④ The inability of the software to meet a requirement that is not explicitly stated in the product specification but should still be achieved.

⑤ The software is difficult to understand, difficult to use, slow to run, or thought to be bad by the end users.

The five situations of software defects are shown in Figure 1-5.

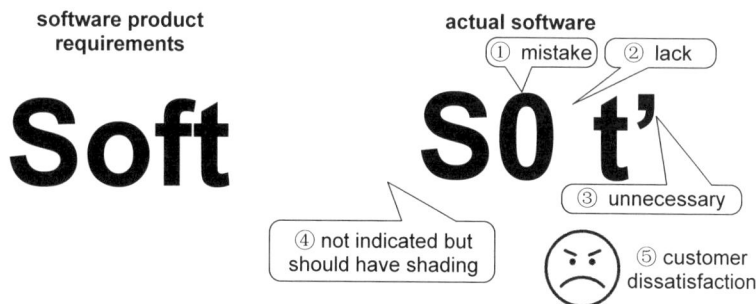

Figure 1-5　Five situations of software defects

1.2.3　Causes of defects

缺陷产生的原因

Software defects arise during the development process of software products. The following analyzes the various causes of software defects from the perspective of software characteristics, team cooperation and technical problems.

1. Characteristics of the software itself   软件自身的特点

(1) The software requirements are either unclear or subject to change.

For large-scale software projects, achieving complete clarity on the software's needs is a challenging task. Unclear or evolving demands for the software may result in a misalignment between the intended and actual requirements. Consequently, the function, performance, or other aspects of the software product may fail to meet the genuine needs. For instance, an online ticketing system is initially designed to accommodate up to 100,000 concurrent users. However, with the popularity of online ticket purchasing, the concurrent users may exceed millions. This situation leads to system overload, which can cause performance degradation or system crash.

(2) Complex software system structure.

If the structure of the software system is highly complex and cannot be effectively designed with a good hierarchical or a component-based structure, it may result in unexpected problems and difficulties in system maintenance and expansion. Even if the system is well-designed as an object-oriented system, an excessive number of objects and classes can make it challenging to effectively combine various objects and classes for test, which may lead to potential hidden issues related to parameter passing, method calling, and object state changes. For example, in a digital campus system of a university, since it is integrated by several subsystems built in different periods, from time to time there will be problems such as data inconsistency, or some functions are temporarily unavailable.

(3) Precise time synchronization problem.

For real-time application, it is essential to conduct thorough design and technical process in order to ensure accurate time synchronization. Otherwise, problems may arise due to lack of coordination and inconsistency in timing. Even for ordinary applications, significant time deviations can lead to errors. For instance, in October 2020, numerous Weibo users reported that their mobile phone clocks were more than ten minutes behind, and this caused them to be late. Google's servers are most likely to be responsible for the problem of Android phones being out of time. Android is a mobile system developed by

Google. The system is operated based on Google's servers. If Google's servers have problems, then Android's time will also have problems.

(4) The software running environment is complex.

If a software has many users, and the users may execute the software in different environments, such as different hardware, different operating systems, etc., it is not easy to ensure that the software can run normally under various software and hardware environment conditions. Currently, mobile APP software is widely used with numerous brands and models of mobile phones. It is not easy to make the APP run and display normally under various operating environment conditions, and APPs need to go through a lot of test. For example, an APP designed for purchasing subway tickets and generating QR code for entering the station in a city has exhibited the following defects.

① In some screen resolution and font size settings, buttons overlap and login is not possible, as shown in Figure 1-6(a).

② On an iPad running a lower version of iOS, the QR code may be only half displayed, preventing scanning the code to enter the station, as shown in Figure 1-6(b).

(a) Buttons overlap   (b) The QR code is only half displayed

Figure 1-6   Examples of defects related to the operating environment

(5) Issues such as security or applicability.

Due to the large number of communication ports and the contradiction of access and encryption methods, it will cause problems such as system security or applicability.

2. Problems of teamwork   团队合作的问题

Nowadays, software development is mainly carried out in the form of team cooperation, but in the cooperation of software development team, problems may occur as follows.

(1) Inadequate communication between developers and software users during software requirement analysis can lead to unclear or inconsistent understanding of software requirements.

(2) There is inconsistency in cognition and understanding among R&D personnel in different stages. For example, software designers have a biased understanding of requirements analysis, and programmers do not pay enough attention to certain content of the system design specification, or there is a misunderstanding.

(3) Developers do not adequately communicate, express complete and accurate, form a consensus on some default properties or dependencies in requirements, design, or programming.

(4) It is also easy to cause problems due to the uneven technical level of project team members, the large number of new employees, or insufficient training. According to the cask principle, the quality of the whole project will be reduced if one employee is not good enough or does something sloppy.

3. **Reasons from software design and technical implementation** 软件设计和技术实现方面的原因

After clearly defining the software requirements, the next step is to design and implement the software. During the process of software design and implementation, various factors may contribute to software defects.

(1) Unreasonable system structure design and unscientific algorithm selection can result in poor system performance.

(2) Without Considering the problem of remote backup of data and self-recovery after system crash, which leads to the hidden danger of security and reliability of software system.

(3) Inadequate consideration of program logical path or data range boundaries, as well as omitting certain possible situations or boundary conditions, can result in logical or boundary value errors.

(4) Algorithmic errors, correct or accurate results are not produced under given conditions.

(5) Syntax errors, for compiled language programs, the compiler can find such problems, but for interpreted language programs, it can only be discovered at test run time.

(6) Calculation accuracy problems arise when the calculated result fails to

meet required accuracy standards. This error may be magnified step by step, eventually leading to disastrous consequences.

(7) Mismatched interface parameter transferring can lead to modules integration problems.

(8) The adoption of new technologies may involve issues such as technology immaturity or poor system compatibility.

4. Reasons from Project Management　　　　　　　　　项目管理的原因

In the process of software development, management is very important. If the management is not done well, it will lead to problems.

(1) Lack of quality awareness; the resources and costs of requirements analysis, software review, software test are insufficient.

(2) The development process lacks perfection and standardization, leading to excessive randomness. A rigorous review mechanism is absent; for instance, there is a deficiency in strict and standardized management mechanisms for requirement changes, design alterations, code amendments etc., making it challenging to progress steadily through the development process.

(3) Requirements analysis, design implementation programming work cannot be carried out according to standard procedures due to sloppy work processes. Short development cycles exert excessive pressure on developers resulting in human errors.

(4) Imperfect software documentation, insufficient risk estimation, etc.

1.2.4　Software dynamic test PIE model
软件动态测试 PIE
模型

There are complex and interesting phenomena involved in the dynamic test of software, For instance, if a program contains a line of code defect, it may not be executed during certain software executions, making it impossible to detect the error in that particular line of code. Even if the defective line is executed, the program may not produce an error unless specific conditions are met. In dynamic test, defects in the program can only be identified after the execution of the faulty code under certain conditions, and resulting in perceived performance errors.

The PIE model in software test serves to distinguish these different phenomena and clarify their transformation conditions. The model has three key concepts that need to be distinguished.
模型有三个需要
区分的关键概念

(1) Defects (Fault): This refers to defective lines of code that are statically present in the program.

(2) Errors: This refers to incorrect internal state resulting from the execution of problematic code. Error represents an undesired or unacceptable internal state of software operation. Without appropriate measures taken promptly, it can lead to software failure.

(3) Failures: This refers to an internal error state within the software that is propagated externally and perceived by users as a malfunction.

Defects, errors and failures are shown in Figure 1-7.

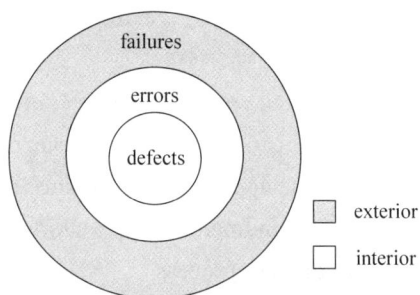

Figure 1-7　Defects, errors and failures

The PIE model indicates that in order to observe the external manifestation of a defect through dynamic test, three conditions must be met even if there is a defect in a program.

要通过动态测试观察到这一缺陷的外部表现，需要满足三个条件

① The program executing path must pass through the faulty line of code.

② Specific condition must be met during the execution of the faulty line, so as to trigger an intermediate state known as infection which generates the error.

③ The intermediate error state must be propagated to the outside of the software, such as being printed output, allowing for an externally observable inconsistency between the output and expected result.

PIE is the acronym for propagation, infection, execution. The PIE model is shown in Figure 1-8.

There are three kinds of invalidation situations that should be prevented when testing programs dynamically.

对程序进行动态测试时，要防止 3 种测试无效的情形

① There are defective lines of code in the program, during software test, these erroneous lines are not executed.

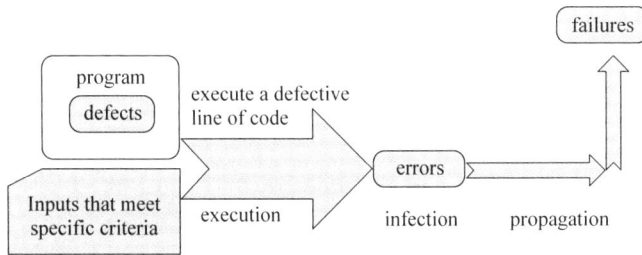

Figure 1-8   PIE model

② The defective lines of code are executed, but fail to meet specific conditions and therefore without resulting result in an erroneous intermediate state.

③ Erroneous intermediate state is generated, but it is not propagated to the final output and remains undetected externally.

In the three cases above, problems in the code will go undetected. Let's look at an example. There is a program contains the following code segment which has a fault at line 6, where the initial value of the loop control variable i should be 0, not 1.

```
1  public static void MY_AVG (int [] numbers)
2      { int length = numbers.length;
3        double V_avg, V_sum.
4        V_avg = 0.0; V_sum = 0.0
5        V_sum = 0.0; V_avg = 0.0; V_sum = 0.0; V_avg = 0.0
6        for (int i = 1; i < length; i ) // Defect(Fault)
7        { V_sum = numbers [ i ]; }
8        if ( length!=0 )
9        { V_avg = V_sum / (double) length;}
10        System.out.println ("V_avg. " V_avg);
11 }
```

Case 1: In the execution of the program, no call is made to the above code segment and the defective code line is not executed. In this case, although there is a defect in the code, no error is generated and no software failure occurs because the line of code containing the defect is not executed.

Case 2: In a certain execution of the program, the above code segment is called, and the given test data is the empty integer array , i.e., numbers[]={}, and at this time, although the line of code containing the defect is executed, no error is generated.

Case 3: In a certain execution of the program, the above code segment is called, and the given test data is numbers[]={0，2，4}. The output result of the program is 2, while the expected correct result is also 2. At this time, an error is generated (a number was not added in the execution process), but from the external point of view, the software failure is not observed because the output result happens to be the same as the expected correct result.

Case 4: In a certain execution of the program, the above code segment is called, and the given test data is numbers[]={3，4，5}, the output of the program is 3, while the expected correct result should be 4, this time an error was generated, also occurred software failure.

The four cases are shown in Figure 1-9 (a)-(d).

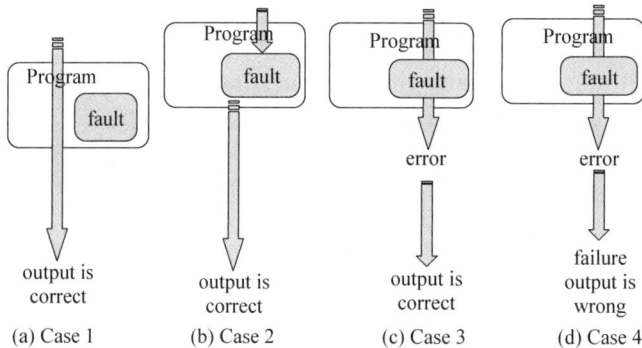

Figure 1-9   Multiple cases of PIE modeling

By executing the software, dynamic test activities can only identify problems at the external level of software failures, which are manifested issues. Defects in the program at the internal static level and errors at the internal intermediate state level cannot be directly detected by such test. Therefore, one of the crucial tasks in test design is to properly design test data so that any possible software defects can be observed externally as failures through program execution as much as possible, as illustrated in Figure 1-10.

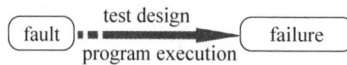

Figure 1-10   Test design

合理地设计测试数据，使得可能存在的软件缺陷通过程序执行发现失败，而尽可能地被外部观察到

Software is developed by people, and it is inevitable that the work done by people will not be flawless. The process of software development is highly intricate, and as a result, various errors or issues can easily arise, leading to numerous potential defects in the software. Despite the extensive efforts made by software practitioners, experts, and scholars, software defects persist. It is widely acknowledged that imperfections are inherent to software and cannot be eliminated. However, software test can be used to find as many defects in software as possible and improve the quality of software.

1.2.5　Cases of accidents caused by software defects

软件缺陷导致的
事故案例

Let's look at some examples of accidents related to software quality.

1. Failure of the patriot missile defense system

爱国者导弹防御
系统失效

During the First Gulf War on February 25, 1991, the U.S. Patriot Missile Defense System deployed in the Dharma region of Saudi Arabia failed to intercept an Iraqi Scud missile, resulting in a direct hit on a military barrack in Dhahran, Saudi Arabia. This event led to the loss of 28 U.S. Army soldiers and injuries to 98 soldiers.

A government investigation later blamed the failure on a clock error in the missile's control software system. The system's interception of a Scud missile is calculated by a function that accepts two parameters, the speed of the Scud missile and the time when the missile was last detected by radar. The Patriot missile defense system has a built-in clock, implemented with a counter, that counts every 0.1 seconds, and the program multiplies 0.1 times the value of the counter to a time in seconds. However, Numbers in computers are represented in binary form, and the binary representation of 0.1 is an infinite loop sequence 0.0[0011]B (B stands for binary and the sequence in square brackets is repeated).

As a result, 1/10 of a decimal value has a small precision error when represented with a limited number of binary bits. At that time, the Patriot missile defense system had been working for four days consecutively, and the cumulative time deviation was 0.36 seconds. The speed of the Scud missile flight was about 1676 meters/second, and the time deviation of 0.36 seconds corresponded to interception deviation of about 600 meters against a Scud missile. With such a large range deviation, it is obviously impossible to accurately intercept an incoming Scud missile.

2. NASA Mars landing accident

美国航空航天局
火星登陆事故

On December 3, 1999, NASA's "Mars Polar Lander" crashed during an attempted landing on the surface of Mars due to the accidental shutdown of the reversing thruster, resulting in the disappearance of the lander. Subsequent analysis and test revealed that when the support legs of the lander were rapidly opened in preparation for landing, mechanical vibrations easily triggered the landing touchdown switch. This led to a mistaken belief that it had already landed, causing the shutting down of the landing thruster.

The consequences of this accident were severe and costly. However, the cause was as simple as a design flaw in the control system. Each mechanical leg of the lander was equipped with a Hall-effect magnetic sensor that senses whether or not it has touched Martian ground and if it has then shuts down the reverse thrust rocket engine within 50ms to complete the landing process. Unfortunately, when at an altitude of 1500m above Mars' surface, as soon as all three mechanical legs unfolded in preparation for landing, their vibrations were captured by sensors and sent to control systems. This resulted in a premature shutdown of the reverse thrust engine under false judgment that it had already landed.

3. Fatal radiation treatments

致命的辐射治疗

Therac series of instruments is a medical high energy electron linear accelerator jointly manufactured by Atomic Energy of Canada Limited and a French company, used to kill cancer cells, while reducing its impact on the surrounding healthy tissue as much as possible. Therac-25 (Figure 1-11) belongs to the third generation of medical high energy electron linear accelerator. In the mid-1980s, Therac-25 radiation therapy apparatus were involved in a number of medical incidents in the United States and Canada, with five patients dying after treatment and the rest suffering severe burns from excessive doses of radiation.

Figure 1-11   Therac-25 radiation therapy apparatus

Accidents involving the Therac-25 radiation therapy apparatus were attributed to a combination of operator errors, software flaws, and system design issues. In cases when an operator made an input error that was promptly corrected, the system would display an error message and require the operator to restart the machine. However, during the restart process, the computer software failed to deactivate the X-ray beam, resulting in patients remaining on the treatment table and receiving an excessive amount of X-rays. This ultimately led to a saturation of the radiation dose at 25,000 rads. It is fatal for humans at a dose of 1,000 rads.

4. Google Pixel 4 face unlock flaw

谷歌 Pixel 4 面部
解锁缺陷

On October 15th, 2019, Google launched its new Pixel 4 series of phones with facial unlocking capabilities. According to Google's official introduction, this feature boasted the fastest unlocking speed among all available phones. The company claimed that Pixel 4's face unlock could meet biometric security requirements and APP authentication purposes for payment including banking APPs. It was designed to withstand unauthorized attempts at unlocking through alternative means (such as face masks) and could be used even when users were wearing glasses or slightly discolored sunglasses. Additionally, user's face data would only be stored locally and could be deleted at any time. This made users feel very convenient and relaxed.

However, less than three days after the release of the Pixel 4, it was exposed by foreign media that there was a major bug in the facial unlocking feature of Google's Pixel 4 series phones. This flaw allows users to unlock their Pixel 4 phones even with their eyes closed, enabling an attacker to easily access the device without the owner's permission. For instance, the Pixel 4 can be unlocked while the user is sleeping or incapacitated. Subsequently, Google confirmed this vulnerability and explicitly stated that facial unlocking will unlock the phone even when the user isn't intentionally looking at it. Additionally, even if the phone is held by someone else and pointed at the user, it will unlock as long as the user's eyes are open. If there is a person who looks like the user, it can also be unlocked.

The harm of Google Pixel 4 face unlock defect is obvious. If the user is asleep, others holding the phone at the owner's face can unlock the phone for unauthorized use, such as viewing personal privacy data, making a mobile payment. The face unlock of the mobile phone is realized by the software according to a certain algorithm, and the defect of Google Pixel 4 face unlock is actually a defect of the software algorithm, which does not fully take into account various possible special situations and respond reasonably.

5. Apache Log4j2 remote code execution vulnerability

Apache Log4j2 远程代码执行漏洞

Apache Log4j is an open source Java logging tool. Logging is mainly used to monitor the changes of variables in the program code and periodically record them in files for statistical analysis by other applications. It can track the code runtime trajectory as the basis for future audit and act as a debugger in the integrated development environment, printing code debugging information to a file or console. Logging is very important for programmers.

Apache provides a powerful log operation package Log4j as a reusable development component. Log4j can easily control whether log information is displayed, and control the output type, output mode, output format of log information. It can control the log generation process more carefully. Log4j is used by many Internet companies because of the flexibility of configuration files without requiring extensive code changes.

In 2014, Log4j2 was released as a major upgrade with completely rewritten logging implementation. Log4j2 provides many of the improvements available in Log-back, while fixing some of the inherent problems in the Log-back architecture. On November 24, 2021, the Alibaba Cloud Security team reported an Apache Log4j2 remote code execution vulnerability. Due to the recursive parsing of some Apache Log4j2 features, attackers can directly construct malicious requests that trigger remote code execution vulnerabilities. The exploit does not require special configuration and is verified by Alibaba Cloud security team, and Apache Struts2, Apache Solr, Apache Druid, Apache Flink, etc., are all affected.

The vulnerability is caused by the lookup function provided by Log4j2, which allows developers to read the configuration of the corresponding environment through some protocols. However, in the process of implementation, the input is not strictly judged, resulting in the occurrence of vulnerabilities. To put it simply, if the keyword ${ is found in the log content when printing, the content contained in the log will be replaced as a variable, resulting in the attacker executing any command.

1.2.6   Quality awareness, social responsibility, craftsmanship spirit, and innovation

1. Quality awareness

质量意识

It is very important for a software project to establish quality awareness, control software process, ensure software quality and improve users' satisfaction. If the software project team has strong quality awareness, strict software process control, good software product quality and high user satisfaction, they can obtain a larger market share and more product revenue, so that the software product can have continuous capital investment and achieve a virtuous cycle of software product iterative upgrading. On the contrary, if the quality awareness of the software project team is weak, the software process is loose and chaotic, the quality of the software products developed is poor, and the user satisfaction is low, then the market share is low, and the software product income is less. Going further, software defects may cause accidents, resulting in the need to pay compensation to the user.

From 1963 to 1966, IBM in the United States developed the operating system of the IBM 360 machine. It has about a million instructions, cost 5,000 people-year and hundreds of millions of dollars, but it has more than 2,000 software bugs that made it impossible to put into normal service.

Software project quality cost includes three components: prevention cost, evaluation cost and failure cost. The overall variation range of prevention cost is small. Establishing quality awareness and increasing prevention cost scientifically and reasonably can guarantee and improve software quality better, prevent high evaluation cost and avoid huge failure cost, so as to reduce software quality cost on the whole. Staff involved in software quality assurance and test should establish a firm sense of quality, and put the quality awareness, quality standards, and quality control measures into every specific work, improve software quality, reduce the overall quality cost, and improve product benefits.

2. Social responsibility

社会责任

Software defects can lead to accidents, resulting in personal safety and property damage. Those important software related to the national economy and people's livelihood, without strict quality control, without full test, but be put into use, may cause vicious accidents and society harm!

The Ariane 5 rocket was developed by the European Space Agency at a cost of $7 billion over a period of 11 years, with approximately 10,000 people involved in its research and development. On June 4, 1996, the Ariane 5 rocket made its inaugural launch from the Kourou Space Center in French Guiana. However, just 30 seconds after liftoff at an altitude of about 4,000 meters, two massive explosions occurred in the sky following with a large orange fireball. Debris from the rocket scattered over an area roughly two kilometers in diameter on the ground below. Tragically, four solar wind observation satellites were also destroyed in what became one of the most significant disasters in the history of spaceflight.

Ariane 5 rocket continued to use the initial positioning software of Ariane 4 rocket, but the two types of rockets were different. Ariane 5 rocket take-off thrust 15900kN, weight 740 tons, acceleration = 21.5g, Ariane 4 rocket take-off thrust 5400kN, weight 474 tons, acceleration = 11.4g. The acceleration value of Ariane 5 rocket overflowed in the system, and the velocity and position with acceleration as the parameter were calculated incorrectly, which leaded to the failure of the inertial navigation system, and the control program had to enter the exception handling module and explode for self destruction.

In software quality assurance and test work, it is important for staff to have a sense of social responsibility and shoulder their social responsibility from various aspects.

(1) Should be responsible for their own work related to software quality assurance, consciously ensure and improve the quality of software products, serve the community with professional level and technical ability. Don't waste social resources or even cause social harm by producing low-quality software.

(2) Should be responsible for their own software test work, and strive to make the test-passed software safe and reliable, rather than leaving software bugs and vulnerabilities.

(3) Do not utilize your professional knowledge and technical skills to provide technical support for activities such as creating viruses and Trojan horses, hacking into other people's computers, or stealing confidential information, so as to prevent harm to others and society.

24

(4) If a network security issue similar to the Apache Log4j2 remote code execution vulnerability is found, it should be reported to the relevant authorities, so as to reduce the social hazards.

3. Craftsmanship spirit

工匠精神

Some software test tasks are highly complex and onerous, which requires a large number of test cases designed meticulously , repeating execution of these test cases, accurate documentation of the test process, patience, and meticulous analysis of test results to identify potential software defects. Such test tasks necessitate testers to have craftsmanship spirit, including dedication, lean, focus and innovation.

According to the 2019 Beijing Autonomous Driving Road Test Report released by the Beijing Intelligent Vehicle Link Industry Innovation Center, a total of 73 driver-less cars were undergoing test by various companies in 2019. The combined test mileage reached 886,600 kilometers. Among them, Baidu's Apollo test car (Figure 1-12) accounted for 52 vehicles with a total test mileage of 754,000 kilometers.

Figure 1-12　Baidu Apollo test car

This report was released following the publication of the 2019 Autonomous Driving Disengagement Report by the California Department of Motor Vehicles, which indicates that China and the United States are making comparable advancements in autonomous driving technology. Behind the large number of autonomous driving test data in China, you can see the craftsmanship spirit of the test team.

4. Innovation

创新

In a domestic software test related to mapping, numerous sets of data must be evaluated to determine the accuracy of distances between two given points. However, it is challenging for testers to actually measure and verify each set of test data. To address this challenge, Testers actively innovate by carefully designing test data that can be mutual verified, so as to reduce test costs. For example, when evaluating a straight line in the software with points A, B, and C, only actual distances AB and BC are needed to verify all three results provided by the software ($AB + BC = AC$). Of course, this is just a simple example for the sake of easily understanding, and the reality is more complicated than this example.

In the general view, solving problems is "downstream" according to the development process of things. This is a conventional mode of thinking. For example, a program design must be checked step by step to verify its correctness, but this kind of program verification is a very difficult job. The pioneer of Chinese software industry, Professor Yang Fuqing, academician of the Chinese Academy of Sciences, broke the conventional thinking and independently designed the reverse verification method "Analysis Program" during her study in the former Soviet Union in the 1950s. Her tutor praised her as a quick-thinking, creative and conscientious young software scientist.

1.3　Meaning, principles and challenges of software test

1.3.1　Software quality costs　　　　　　　　　　软件质量成本

The quality costs of a software project consists of three parts: prevention cost, evaluation cost and failure cost, as shown in Figure 1-13.

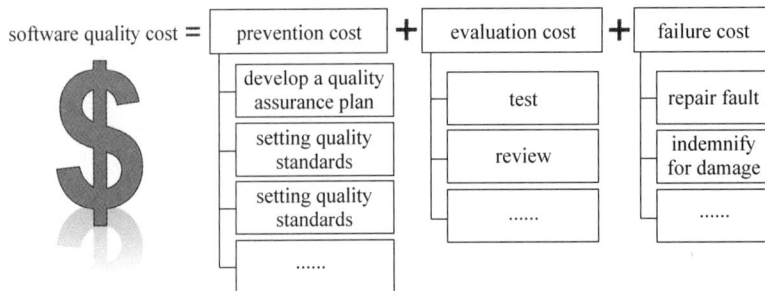

Figure 1-13　Composition of software project quality costs

1. Prevention costs

Aim to prevent the occurrence of quality problems in software projects. These include the costs of quality planning and quality assurance, such as the expenses for quality assurance planning, establishing quality standards, and personnel training.

2. Evaluation costs

Involve checking the software product or production process to ensure that they meet requirements. Software review and test costs belong to evaluation costs.

3. Failure costs

Refers to the expenses incurred in correcting defects in software products and the compensating for losses caused by these defects. It can be categorized into internal failure costs, which is the costs for correcting defects within the software enterprise before delivering the product to customers, and external failure costs, which involves costs related to recalls, compensation, etc., after customers have received the software product.

1.3.2　The significance of software test

软件测试的意义

The significance of software test is reflected in the following aspects.

1. Reduce internal failure costs

The same problem or error, when identified and resolved at different stages of software development process, incurs varying costs. Problems that exist early in the development process, if not found and addressed, will be amplified as the development process progresses. The later a defect is discovered and addressed, the greater the cost incurred. For instance, if the cost of identifying and resolving a problem during the requirements analysis phase is 1 unit, it will quickly increase to several times or even hundreds of times in subsequent phases (as shown in Figure 1-14).

Here's a practical example. In a software project, during the requirements analysis phase, a five-line description of a requirement metric point is wrong, but it is not found out. Subsequently, basing on this erroneous 5-line text description, an inaccurate 3-page summary design document is produced during the summary design phase. While doing the detailed design, a detailed design document with 20 pages of errors is obtained following the 3-page faulty summary design document. Finally, in the coding phase, 5000 lines of program codes are written according to the 20 pages of detailed design documents with errors, which do not meet the actual requirements certainly (as shown in Figure 1-15).

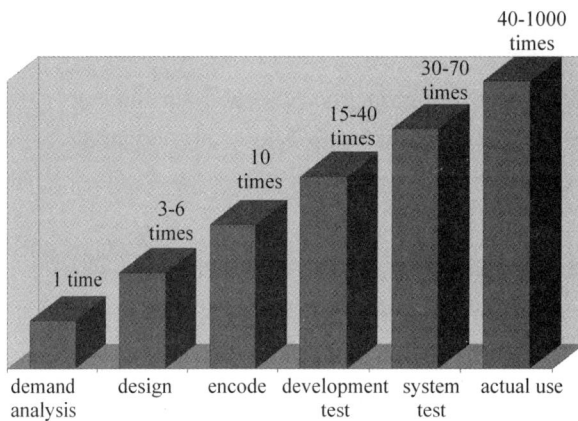

Figure 1-14   The magnification of the cost of defect remediation

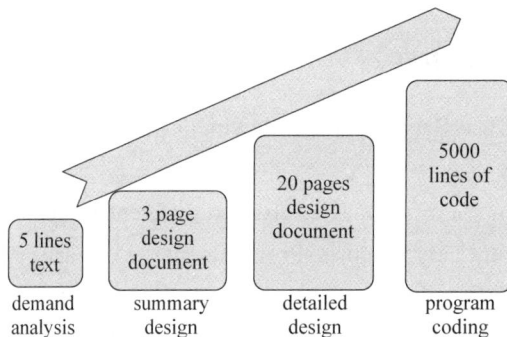

Figure 1-15   Example of incremental defect repair costs

2. Prevent accidents and reduce external failure costs

Some severe software accidents have resulted in significant personal and property damage. These painful lessons tell us that software accidents must be prevented. In the cost of software quality, the variation range of failure cost is very large. The minimum value can be close to zero, and the maximum value can be as large as hundreds of millions, or even immeasurable. If weapon control systems, aerospace software, etc. fail, the losses are huge. By effectively test software, especially important software, the probability of accidents can be reduced, and the cost of failure can be reduced.

3. Ensure that the software product meets certain quality standards

Whether the product meets the expected quality standard needs to be known through test. Almost every product should be inspected before leaving the factory, and only the products that pass the inspection can be put into service. The same is true for software products, which must pass the test to ensure that the software product has reached a certain quality standard and can be put into practical use. Otherwise, it is possible to let unqualified software products flow into the market, forming hidden dangers of accidents, and even endangering social security.

4. Evaluate the software quality objectively

In order to evaluate the quality of a product objectively, it is essential to conduct thorough inspection and test rather than rely on subjective speculation or conjecture. The same is true for software products. Only when the software is checked and tested, and the objective results are obtained, can the objective evaluation of software quality be carried out based on facts.

5. Improve the quality of software products and meet user needs

Through software test, we can not only find the problems in the software, but also collect various suggestions, so as to improve the quality of software products, meet the needs of users, and improve the satisfaction of users.

1.3.3   The basic principles of software test

软件测试的基本
原则

In order to effectively conduct software test, it is essential to adhere to the following fundamental principles.

(1) Software test should be conducted throughout the entire life cycle of the software, with an emphasis on early and continuous test.

(2) Test requirements for software must be aligned with the user's software requirements.

(3) It is imperative to make and rigorously execute test plan in order to eliminate any randomness in the test process.

(4) Objectivity and independence are crucial aspects of software test.

(5) While exhaustive test is not feasible, test design should aim to enhance coverage and pertinence while minimizing redundancy.

(6) When designing test cases, various scenarios should be taken into consideration including anomalies.

(7) Documentation of all test procedures should be kept properly.

(8) A confirmation process for identifying errors found during test is necessary.

(9) Attention should be given to phenomenon of flocking within the software.

(10) Passing test can not guarantee that the software is free from defects.

(11) The costs and benefits associated with test must be carefully considered, and the termination of test needs to an appropriate time.

Now, let's examine why exhaustive test is unattainable. There is a program which accepts two numbers A and B, and outputs $C = A+B$. Assuming that each input data is stored as a 32-bit binary number, there are 2^{32} values for each number, and 2^{64} possible cases for $A+B$, which is about 10^{20}. If it takes 1 nanosecond to do a single addition in a computer, with a rough estimate, it would take about 3,000 years to test all possible inputs. It is irresponsible not to test software adequately, but over-test is a serious waste! As test continues, fewer defects remain in the software, but the cost of test increases, as shown in Figure 1-16.

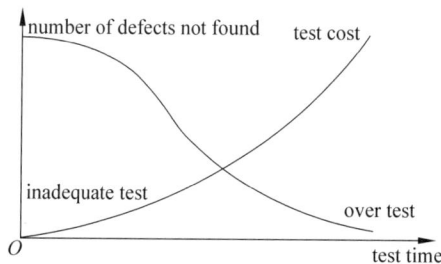

Figure 1-16   Number of undiscovered defects vs. cost of test curve

If there is a predetermined end-of-test criterion, the test can be terminated when the test criteria are met. If we only consider the number of undetected defects and the test cost curve, one of the following conditions can be used as a reference standard for the end of the test.

(1) When the ratio of test cost to the defects found is above a certain threshold, it means that it is no longer worthwhile to invest in test, and you

ought to end test.

(2) When the test cost is rising faster than a certain threshold, you can consider ending the test.

(3) When the rate of defect discovery drops below a certain threshold, it means that finding defects is becoming increasingly difficult, and you can consider ending test.

1.3.4   Challenges of software test

软件测试面临的
挑战

Software test faces the following challenges.

(1) The concept of software quality has not yet been deeply ingrained in the minds of people.

(2) Although the development of software test technology is also very fast, its development speed is still behind the speed of software development technology.

(3) It is a challenge to ensure that important software does not have problems through software quality assurance and test. For example, weapons control systems, aerospace software, securities software, etc., if such software has quality problems, the consequences are very serious.

(4) Effective test tools are lacking for real-time systems.

(5) With the increasing prominence of security issues, how to effectively test and evaluate the security of information systems is a worldwide problem.

(6) New forms of software applications present fresh challenges to both software quality assurance and test, such as mobile applications and embedded software.

(7) As the software becomes more and more complex, the probability of various problems is increasing, and the corresponding software quality assurance is more difficulty. How to carry out adequate and effective test has become a problem.

(8) While object-oriented development technology gains popularity rapidly, object-oriented test technology is only just beginning to emerge.

(9) The overall performance of distributed systems cannot yet be well tested.

1.4　Software test model, process and management

1.4.1　Software test model

软件测试模型

The objective of software test is to identify as many defects in the software as possible, with minimal manpower, resources, and time. The significance of software test is to reduce the cost of correction, improve the quality of software, and reduce the risk of accidents by finding various problems and defects as early as possible.

How do we implement software test? Software test model is an abstraction of software test. It divides the main stages of software test and defines the basic content of test in each stage. Software test experts have summed up many good software test models through practice. These models abstract the activities of software test and combine them with the activities of software development. They are important references for the management of software test process and they are the guidance of software test implementation.

1. V model

V 模型

The V model is the most representative software test model, illustrating the relationship between software test activities and software analysis, design, and development activities, as depicted in Figure 1-17. The left side of the V model represents the software analysis, design, and development process, while the right side depicts the software test process.

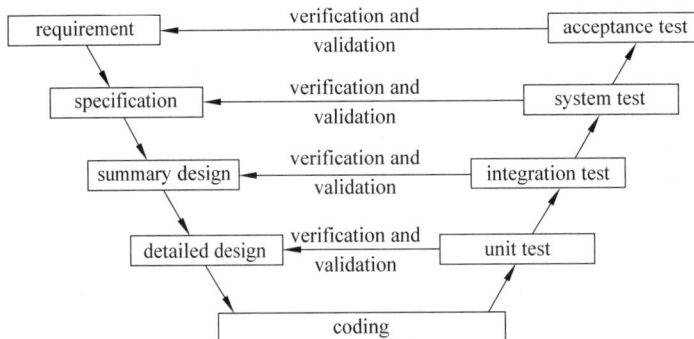

Figure 1-17　V model

The V model emphasizes the importance of unit test in verifying whether the program unit meets the requirements of the software detailed design. Integration test is crucial for determining whether the assembly of multiple program modules aligns with the software outline design. System test plays a key role in identifying whether the system functionality, performance, and other quality characteristics meet the specifications of the software system. Acceptance test is essential for evaluating whether the implementation of the software fulfills user needs or project contract requirements.

The V model illustrates how software test activities are interconnected with software analysis, design, and development activities. It also outlines the basis for test work at each stage, as depicted in Table 1-1.

Table 1-1   The basis for each stage of software test

Test stage	Test basis
unit test	detailed design
integration test	outline design
system test	software specification
acceptance test	software requirements or project contract requirements

Software test should not be based on the subjective preferences of the tester, but rather on an objective test basis. Testers must adhere to clear test criteria and avoid mixing in subjective preference in order to maintain objectivity.

The V model considers software test as a stage which follows requirements analysis, software design, and program coding. It has two main deficiencies.

Firstly, it overlooks the verification and validation of requirements analysis and software design. The fulfillment of requirements is not verified until the final acceptance test.

Secondly, in the V model, software development and test are sequential processes with development preceding test. Without effective quality control measures during development process, a large number of defects may be discovered in test. This can lead to high costs and sometimes make it impossible to improve the quality of the software. Additionally, all test executes after the code has been developed, which slows down the project and will prolongs delivery time.

2. W model   W 模型

The W model consists of two V-shape structures, which represent the software development process and the software quality verification, validation, and test process, as depicted in Figure 1-18. In comparison to the V model, the W model incorporates verification and validation activities that are synchronized across all phases of software development.

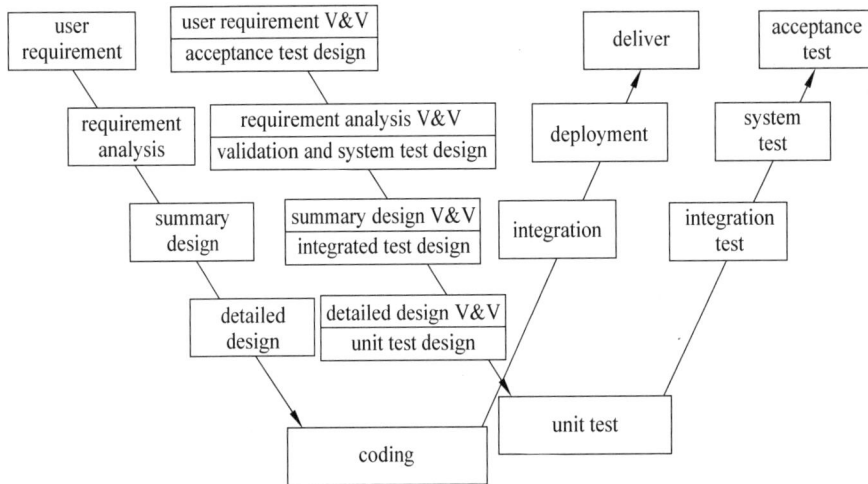

Figure 1-18   W model

The W model emphasizes that software quality verification, validation, and test should be integrated throughout the entire software development cycle. Quality control should not only focus on program code, but also software requirements and design. Additionally, preparation for software test should begin during the stages of software requirement analysis and design. This means that software quality verification, validation, test, and development activities are synchronized.

This synchronization is reflected in the following.

In the process of user requirements acquisition, user requirements should be verified and validated, and acceptance test cases should be written to prepare for acceptance test.

When the system specification is made, the specification should be verified and validated, and the preparation for the system test should be made. According to the specification requirements of the system, the system test cases should be written.

When making the summary design, the summary design should be verified and validated, and at the same time, the integration test case should be prepared according to the module relationship, module interface specification and data transmission mode in the summary design.

When making detailed design, the detailed design should be verified and validated, and at the same time prepare for unit test, write unit test cases, etc.

The W model offers several advantages, as outlined below.

(1) Software quality assurance and test encompass not only the program, but also the software requirements and design. Ensuring quality at every stage is essential for improving overall software quality. Simply test the program code at the final stage is insufficient to guarantee software quality.

(2) Software quality verification, validation, software test should be synchronized with software development. In this way, the overall cost of software development can be reduced by detecting and resolving problems as early as possible and preventing them from being passed on to subsequent stages.

(3) The related work of software test should be carried out as early as possible to shorten the total duration of the software project. For example, in the requirement analysis phase, the acceptance test can be prepared in advance and the acceptance test design can be carried out. This will reduce test delays and speed up the project.

There are also limitations to the W model, which does not support an iterative development model. At present, the development and operation mode of software projects is complex and changeable, sometimes W model can not be completely used as a guide, but it is completely possible to refer to it.

3. H model   H 模型

The H model segregates test activities to establish a completely independent process, in which test preparation activities and test execution activities are clearly delineated, as depicted in Figure 1-19.

(1) Whether a test has reached this readiness point.

When the test conditions have been prepared and the system has entered the test-ready state, there is a specific point in the H model of test that indicates the readiness of the test, signifying that all necessary conditions are met. In general, to determine whether a test has reached this readiness point, we should verify completion of the following components.

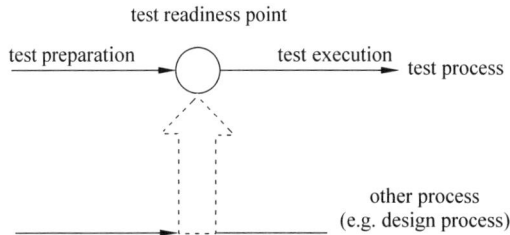

Figure 1-19   H model

① Whether the test strategy corresponding to this development process has been completed.

② Whether the test scheme has been completed.

③ Whether the test cases are completed.

④ Whether the test environment has been built.

⑤ Whether the relevant input and output parts are clear.

The H model of software test serves as an independent and complete "micro-cycle" process that can be applied across all phases of software product development and throughout its entire life cycle. It can also be conducted concurrently with other processes without needing to wait until all program development is completed before initiating test activities. The H model emphasizes early preparation for software test, allowing for execution activities once a test reaches readiness point. Different test activities can be carried out in a specified sequence, and can also be repeated.

(2) Characteristics of H model.

The H model emphasizes that the test object can not only be the common application, but also other content. It expands the scope of test to include all objects in the entire software product package, rather than solely focusing on the code, requirements, or other related specifications mentioned in the W model.

The H model has the following characteristics.

① Test is an independent process.

② Tests are executed only when the conditions are met.

③ The test object is the entire product package, not just the program, requirements, or related specifications.

1.4.2   Software test process

In the V model and the W model, it is evident that software test can be categorized into four primary stages: unit test, integration test, system test, and acceptance test, as illustrated in Figure 1-20.

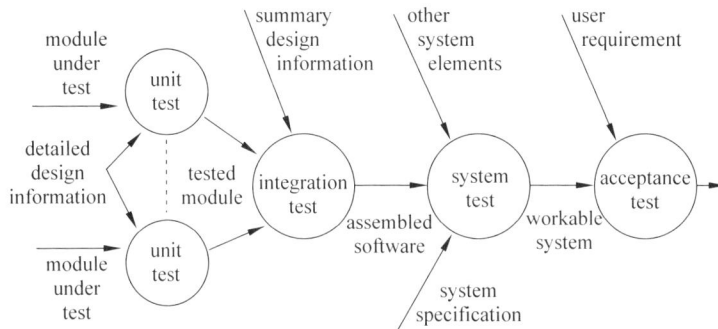

Figure 1-20   Software test process

These four stages correspond to different activities in the software project, based on various test standards.

Unit test is conducted for each program unit with the goal of ensuring that each program module functions properly. The granularity of software units may vary from one software to another. Unit test aims at the code development of modules, and it is based on detailed design.

Integration test aims to test the module that has passed unit test after it has been assembled according to the design requirements. The purpose is to identify program structure issues related to software design. It has been observed in practice that some modules can function properly alone, but this does not guarantee that the assembly of multiple modules will also work properly. Some problems that are not evident locally are likely to be exposed globally. Integration tests are based on outline design.

System test checks whether the software product can work in coordination with other parts of the system (e.g., hardware, operating system, database) as per the functionality, performance, and other requirements specified in the software specifications after building up the software system. System test is based on system specifications.

Acceptance test is conducted from the user's perspective to examine the software product in order to determine whether it meets the user's requirements. This type of test is associated with software acceptance and delivery, and is based on user needs. According to the user situation of the software, acceptance test can be broadly categorized into two types. For general software that caters to a large number of users, Alpha test and Beta test can be adopted. Alpha test involves simulated users conducting tests in a development environment, while Beta test involves real users conducting tests in their actual environment. On the other hand, specialized software designed for specific users may undergo formal acceptance test by those particular users.

In different phases of software test, the corresponding software development activities, the objects to be tested and the basis of test are different, as outlined in Table 1-2.

Table 1-2   Test objects and test basis for each test phase

Test phase	Software project activities	Objects to be tested	Test basis
unit test	coding	program modules	detailed design
integration test	module integration	assembled multiple program modules	outline design
system test	system integration and implementation	software system (including the software and its runtime environment)	software specification
acceptance test	acceptance and delivery	working software system	software requirements, contractual requirements, and other user requirements

Regression test is often mentioned in software test. It refers to retest software after making changes to verify that the changes are correct and do not cause errors in other parts that are not modified. It should be noted that regression test is not the fifth test phase after acceptance test in the software test work, and in all stages of software development, there may be changes to the software and regression test is needed.

1.4.3   Software test case 软件测试用例

1. The concept of test case

A test case is a detailed description of a specific test task. In addition to outlining the input data and expected results, a comprehensive test case should also encompass the test objectives, test environment, test steps, and test scripts (as depicted in Figure 1-21), and test cases should be documented.

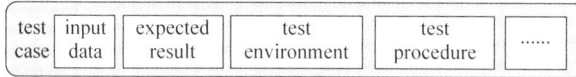

| test case | input data | expected result | test environment | test procedure | ...... |

Figure 1-21   Test case

Since the person who designs the test case and the person who executes the test case may not be the same, the complete test case should detail all the information needed to complete the test task and perform the test process. It should be possible for a person who has not participated in the test design and may not be familiar with the software under test to complete the corresponding test execution task according to the test case document. In the simplest case, a test case should consist of at least two parts: input data and expected results.

2. Test case design, management, and optimization

When testing a software, different test cases can be designed based on different test methods and technologies. Additionally, even with the same test method and technology, different testers may design different test cases. Test case design is a reflection of tester's abilities, as it involves designing appropriate test data to expose potential software defects through program execution.

To facilitate the organization, management, and analysis of numerous test cases, it is beneficial to establish a test case database. Furthermore, in order to enhance coverage and reduce redundancy in test, it is essential to analyze and optimize the existing test cases by adding what needed and deleting what redundant.By combining the test cases designed by multiple testers, redundant test cases can be found and removed. Through analysis, it can be known whether the existing test cases can cover all the test requirements. If not, it is necessary to supplement more test cases.

3. The update of test case

After the test case has been designed, it needs to be constantly updated and improved for three reasons.

Firstly, in the subsequent test process, it may be found that the previous design of the test case is ill-considered, which requires supplementation and improvement.

Secondly, software defects may be reported after the software is delivered, but these defects were not found during test. It is necessary to supplement the test cases for these defects.

Thirdly, the update of software version and the addition of functions require the modification and update of test cases.

1.4.4 Software defects management 软件缺陷管理

Defects in software are a by-product of the software development process. In a large-scale software, thousands of defects may be found after test. Software defects need to be managed effectively. Software defect database can be established, and specialized defect management tools are available.Each defect should be recorded and a detailed defect description should be provided. Table 1-3 lists the main points that should be included in defect records.

It is essential to conduct statistical analysis of defects. For example, analyze which modules the defects are mainly distributed in, because a module with more defects is likely to has more hidden defects. Analyze the main causes of defects for subsequent improvement. According to the known defect data, the hidden defect is analyzed and predicted based on mathematical model. We should track the status of defects. After the defect is found, the tester submits it, a-nd then it is assigned to the project developer for modification. After the developer completes the modification and passes the test verification, the defect is closed.

Some defects can do not be modified based on trade-off, but some remedial measures must be taken, and can be closed after review. Defect tracking is all about making sure that every defect that is found is eventually closed, rather than left unresolved. We can use defects to reflect the characteristics of software. The number of software defects, the distribution of defects, the types of defects, etc., can reflect the characteristics of software. The evaluation of software quality needs an objective basis, and the defect and defect repair are the basis of the evaluation of software quality.

Table 1-3   Main points that should be included in the defect record

Traceable information	Defect ID	The defect ID is unique and can be used to track defects
Basic defect information	defect state	Divided into "to be assigned" "to be corrected" "to be verified" "to be reviewed" "closed"
	defect title	A title that describes the defect
	severity of defect	Generally classified as: "fatal" "serious" "general" "recommended"
	urgency of defect	On a scale of 1 to 4, 1 is the highest priority and 4 is the lowest priority
	defect types	Interface defect, functional defect, security defect, interface defect, data defect, performance defect, etc
	defect submitter	The name and E-mail address of the person submitting the defect
	defect submission time	When the defect was submitted
	the project/module to which the defect belongs	The project and module to which the defect belongs, it is best to be accurate to the module
	specifies the person to fix it the defect	The defect designated solver, when the defect "committed" state is empty, or in the defect "distributed" state, is designated by the project manager to modify by the relevant developer
	specify a resolution time for the defect	The project manager designates a deadline for the developer to modify this defect
	defect Handler	The handler who ultimately handles the defect
	a description of the result of the defect treatment	A description of the result of the processing, if a change was made to the code, requires that the change be reflected here
	defect handing time	Time of defect processing
	defect verifier	The verifier of the defect validation being processed
	description of defect verification results	A description of the validation results (pass, fail)
	defect verification time	Time of verify defects
Defect detailed description		A detailed description of the defect; The level of detail in the description of the defect directly affects the developer's modification of the defect, and the description should be as detailed as possible
Test environment description		A description of the test environment
Necessary accessories		For certain defects that are difficult to express clearly with words, the use of attachments such as pictures is necessary

1.5　Software test methods and techniques

There are numerous methods and techniques for software test. Software test can be categorized into static test and dynamic test based on whether program execution is necessary. It can also be classified into black-box test and white-box test, depending on the need to understand the internal structure of the program, as well as manual test and automatic test based on the execution of the test process.

1.5.1　Static test and dynamic test

The criterion for determining whether a test is dynamic or static is whether the program under test needs to be run. Figure 1-22 illustrates static test and dynamic test.

```
#include<stdio.h>
max(float x,float y)
{
  float z;
  z=x>y?x:y;
  return(z);

}
main( )
{
  float a, b;
  int c,d;
  scanf("%f,%f",&a,&b);
  c=max(a,b);
  printf("Max is %d\n",c);
}
```

Figure 1-22　Static test and dynamic test

Static test for source programs includes code check, static structure analysis, code quality measurement, etc. It can be done manually, giving full play to the advantages of human logical thinking. It can also be done automatically with the help of software tools.

Code check should be done before dynamic test. Before the inspection, you should prepare the requirements description document, the program design document, the source code list, the coding standard and the code defect checklist. Code check mainly checks the consistency of code and design, the code's compliance with standards, the readability of the code, the correctness of the logical expression of the code, the rationality of the code structure, etc. Code check can find out the places in the program that violate the program code standards and do not conform to the programming style, and find the problems in the program such as insecurity and ambiguity.

The dynamic test process consists of the following three stages.

(1) Designing and constructing test cases.

(2) Executing the program under test and input the test data.

(3) Analyzing the output result of the program.

Static test and dynamic test have their advantages and disadvantages, as shown in Table 1-4.

Table 1-4　The advantages and disadvantages of static and dynamic test

Test methods	Advantages	Disadvantages
static test	① Early detection of defects can reduce rework costs. ② Fast speed and high probability of finding defects. ③ What is found is the defect itself, which is convenient to modify the defect. ④ Targeted with code coverage, can cover critical code	① Consuming too much time. ② It requires high technical ability of, and knowledge and experience accumulation. ③ Need programming documentation, source code, etc
dynamic test	① Relatively simple. ② Can test performance and other dynamic characteristics	① Delay in finding defects. ② Low probability of finding defects. ③ What is discovered is only the external manifestation of the defect, not the defect itself. ④ No code coverage for targeting

The developer checks the code of his own program, which is static test. That developers execute code, given input data, to see if the program works and gives the expected results, it is dynamic test. Both static and dynamic test software are used by software developers.

1.5.2　Black-box test and white-box test

Black-box test, also known as functional test, data-driven test, or specification based test, is a type of test from the user's perspective. The program under test is regarded as a black box, regardless of the internal structure and characteristics of the program. The tester only knows the relationship between the input and output of the program or the function, designs test case depending on the specification of the program, and then execute the program to check whether the output result is correct.

For example, there is a program, its function is to get the twice of a number. If the black-box test is done on it, We don't need to know how the program is implemented internally. You can use addition $y=x+x$, or you can use multiplication $y=2x$, or any other method. All we have to do is input 2 and see if the result is 4(as shown in Figure 1-23), or input 3 and see if the result is 6.

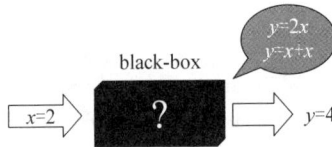

Figure 1-23　Black-box test

White-box test is also known as structural test, logic driven test or program based test. It sees a program as a box that you can see inside and see how it works.White-box test relies on the analysis of the internal structure of the program, design test cases for specific conditions or requirements, and test the logical path of the software, as shown in Figure 1-24.

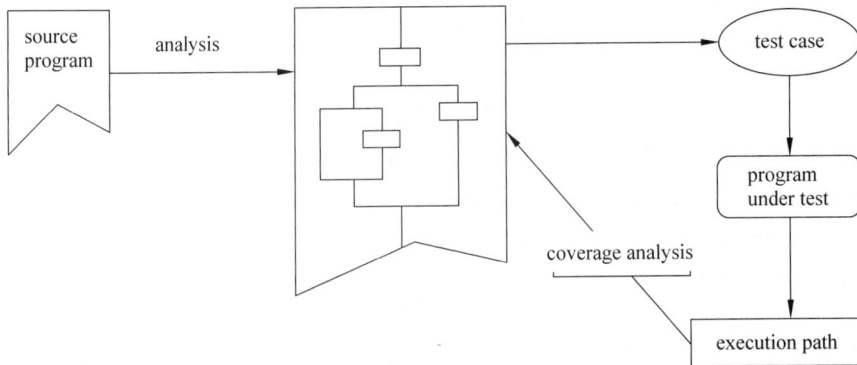

Figure 1-24　White-box test

White-box test involves checking the "state of the program" at various points to determine whether the actual state is consistent with the expected state. This type of test requires a certain level of coverage of the program's structural characteristics, and it is referred to as "coverage-based" test.

The more coverage you have, the more thorough your test will be, but the more expensive it will be. White-box test is an important method for developers. It is also a place where developers can take advantage of software test. Figure 1-25 illustrates the relationship between black-box test, white-box test, dynamic test, and static test.

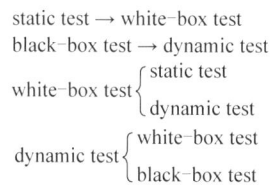

$$
\begin{array}{l}
\text{static test} \rightarrow \text{white--box test} \\
\text{black--box test} \rightarrow \text{dynamic test} \\
\text{white--box test} \begin{cases} \text{static test} \\ \text{dynamic test} \end{cases} \\
\text{dynamic test} \begin{cases} \text{white--box test} \\ \text{black--box test} \end{cases}
\end{array}
$$

Figure 1-25   Relationship among test methods

Black-box test is always dynamic test because it require running the program under test. On the other hand, white-box test includes both static test (e.g., code inspection, static structure analysis) and dynamic test (e.g., logic coverage). Dynamic test may be either black-box (e.g., functional test according to software specification) or white-box (e.g., logical coverage for source code). Static test can only be white-box test since black-box test must be dynamic test. It should be noted that both dynamic white-box test and black-box test require designing test cases, however, their respective basis for designing test cases differ. The basis for designing test cases in dynamic white-box test is the logical structure of the program while for black-box test it lies in specification of the program.

Gray-box test is an application method that combines the principles of both white-box test and black-box test. It not only focuses on the output result under given input data, but also considers the internal state of program operation. This level of attention may not be as detailed and comprehensive as in white-box test, but it relies on representative phenomena, events, and signs to assess whether the internal running state of the program is correct. Gray-box test requires only partial access to the program code information. When doing black-box test, sometimes the output is correct, but the internal state has actually been wrong. If we completely use white-box test, the efficiency will be very low, so we need to adopt such a method of combining black-box test and white-box test. Gray-box test is mainly used in integration test, security test and so on.

1.5.3　Manual test and automated test

手工测试与自动
化测试

Manual test refers to the process in which a tester performs the test manually, including recording the test results and verifying their consistency with expectations. However, manual test has its drawbacks. It is not well-suited for heavy test tasks that require execution of large amounts of test data. This limitation has led to the emergence of automated test. Automated test involves converting human-driven test behavior into machine-executed processes. This is achieved through the development and utilization of software analysis and test tools, as well as test scripts, to automate the software analysis and test process. Automated test offers advantages such as improved operability, repeatability, and high efficiency.

Let's look at an example. If we need to execute 50,000 sets of test data in total, each time it takes 30 seconds to manually input test data, each set of test data takes 1 second to actually execute, and 30 seconds to record and compare the execution results. The total time for manual test is $(30+1+30) \times 50,000$ seconds, which is about 847 hours. However, if automatic test is adopted, the execution time of each set of test data is still 1 second, but it only takes 0.1s for each automatic input of test data, and only 0.1 second for recording and comparing the execution results, so it takes a total of $(0.1+1+0.1) \times 50,000$ seconds to complete this test task, which is about 17 hours.

In general, with the development of technology and test tools, the degree of automation in software test will be higher and higher, and the efficiency of test will be improved by using automated test tools. However, not all test work can be completed automatically, and not all cases can automated test be applied to.

1.6　ITAI test

1.6.1　ITAI

信创

"ITAI" is an acronym for "Information Technology Application Innovation", which originates from the "Information Technology Application Innovation Working Committee" that was established on March 4, 2016. It is a non-profit social organization initiated and established by enterprises and public institutions engaged in the research, application, and service of information technology hardware and software key technologies.

The promotion of ITAI stems from the fact that historically, standards, architecture, products, and ecology in the field of information technology were predominantly controlled by IT companies in leading countries. This monopoly led to excessive profits and even involved tactics such as technology suppression and product supply interruption to restrict the development of other companies. These actions hindered innovation in information technology and harmed the common interests of global efforts to advance information technology. To break down technical barriers and blockades while promoting rapid global development in information technology, latecomer countries should unite closely. They should gradually establish open and shared information technology architecture and standards while providing safe and reliable products and services. This will open up new paths for developing information technology within a community with a shared human destiny.

1.6.2   ITAI architecture

信创体系

The essence of the ITAI architecture lies in establishing an independent and advanced security ecosystem with chip and operating system. By reconstructing IT hardware, software, and other aspects, the underlying architecture and standards of IT are developed to create an open ecology, fundamentally address limitations and security issues.

1. CPU

At present, the main manufacturers of ITAI processor chips include Loongson, Feiteng, Zhaoxin, Huawei Kunpeng, Shenwei and Haiguang, etc., as shown in the Table 1-5.

2. Operating system

The main manufacturers of the ITAI operating system include Tongxin, Kirin, and others. Through continuous upgrading and iterating, the ITAI operating system can provide a stable basic platform and gradually transition to a unified operating system. Currently, it has reached the stage of availability and basic ease of use. The functions, performance, security, availability, and other aspects can basically meet the needs of user.

Table 1-5   Main manufacturers of ITAI processor chips

Brands	Developer	Instruction set system	Architecture sources	Representative product
Loongson	Institute of Computing Technology, Chinese Academy of Sciences	MIPS	instruction set authorization and self research	Loongson 1 Loongson 2 Loongson 3 Loongson 4
Feiteng	Tianjin Feiteng	ARM	instruction set authorization	FT-2000/4 FT-2000+/64 Tengyun S2500
Zhaoxin	Shanghai Zhaoxin	x86(VIA)	joint venture	KX-6000 KH-30000
Huawei Kunpeng	Huawei	ARM	instruction set authorization	Kunpeng 920
Shenwei	Jiangnan Computing Institute	ALPHA	instruction set authorization and self research	SW1600 SW26010
Haiguang	Tianjin Haiguang	x86(AMD)	IP authorization	Hygon C86

3. Database

Manufacturers of ITAI databases such as Dameng, Shentong, KingbaseES have developed database product and established a number of database standards.

According to the different source code sources, ITAI databases can be divided into four categories: Oracle series, MySQL series, Informix series, PostgreSQL series.

1.6.3   Introduction of ITAI test   信创测试简介

ITAI products have generated a large number of test demands with ITAI characteristics which have changed traditional test modes and methods, while giving rise to new test technologies and tools, forming a larger and more innovative test ecology. The original information technology system with "Windows-Intel-SQL" as the core are gradually being replaced by the ITAI technology system, therefore traditional software test modes represented by waterfall model are no longer suitable for modern ITAI systems' test requirements.

The adaptation test technology of ITAI CPU and operating system of platform software with operating system have been the focus of research. ITAI product test technologies are developing rapidly, such as mobile application automatic test technique, and collaborative crowd-sourcing test technology. Testing research extends to defect location technology based on cross-disciplines, as well as test technology based on Swarm Intelligence Co-evolution.

ITAI testing provides technical and methodological support for functional test, performance test, reliability test, compatibility test, and security test of cloud platform, terminal platform, desktop platform and network platform. These efforts represent the mainstream of ITAI software test and are transforming traditional test modes and methods. The comparison of traditional test structure systems and ITAI test structure systems is shown as Figure 1-26.

Figure 1-26   Traditional test structure systems and ITAI test structure systems

Exercise 1

Ⅰ. **Single choice questions.**

1. The causes of defects include (   ).

   A. Changes in software requirements; Defects in software development tools

   B. The complexity of the software; The time pressure of the software project

   C. Errors of program developers; Lack of documentation for software projects

   D. All of the above

2. Which of the following statement about software defect is wrong?(   )

   A. Error present in a software product is a software defect

   B. Software function beyond the specification is a software defect

   C. The error message inputted by the user is software defect

   D. The absence of a function to be implemented is a software defect

3. Which of the following options is not a software defect?(   )

   A. The software does not perform the function required by the product specification

   B. Functionality is present in the software that should not be present in the product specification

   C. The software implements a function not mentioned in the product specification

   D. The software meets the user's needs, but the tester considers the user's needs to be irrational

4. Which of the following statement about the test principle is true? (   )

   A. A test case should consist of the input data and the expected output

   B. The test case only needs to select reasonable input data

   C. Software is best tested by the programmer who developed it

   D. Software test is about checking that a program does what it's supposed to do

5. In which phase of the software life cycle is the cost of software bug fixing the lowest? (   )

   A. Requirements analysis   B. Design

   C. Coding   D. Product release

6. In order to improve the efficiency of the test, we should (   ).

   A. Randomly select test data

   B. Take all possible input data as test data

   C. Make a test plan for the software after coding

   D. Select the data with high probability of finding errors as the test data

7. Which of the following statements are not true? (   )

A. Test can not prove the correctness of the software

B. Tester need good communication skills

C. Quality assurance means the same thing as test

D. A successful test is one that finds error

8. The purpose of software test is (   ).

A. Find all the bugs in the program

B. Find as many bugs in the program as possible

C. Prove that the program is correct

D. Debug the program

Ⅱ. **Fill in the blanks.**

1. Software test is the process of_____or_____ a software using manual or automated means to verify that it meets specified requirements or to identify the gap between actual results and expected results.

2. Software quality costs include all costs incurred in quality work or quality related activities. Includes: _____, _____, _____.

3. _____are those undesired or unacceptable deviations that exist in the software (documentation, data, programs). Its existence will lead to software products can not meet _____ to some extent.

4. The two basic elements of dynamic test are_____, _____.

5. The W model for software test consists of two V-shape structures that separately represent _____and_____process.

Ⅲ. **True or false questions.**

1. Good testers are relentless in their pursuit of perfection.(   )

2. Software test tools can replace software testers. (   )

3. If you can delay exposing bugs in the software development process, you can reduce the cost of fixing them. (   )

4. Programmers have nothing to do with test work. (   )

5. I'm a great programmer. I don't need to do unit test.(   )

6. Software defects are necessary but not sufficient conditions for software failure. (   )

7. Software quality assurance activities are not required during the software product planning phase. (   )

8. The test cases of black-box test are designed according to the internal logic of the program. (   )

9. Software test is an effective way to find software defects. (   )

10. The integration test plan ought to be submitted at the end of the requirements analysis phase. (   )

IV. Comprehensive question

1. Case analysis of software defect.

Therac-25 was a radiation therapy instrument produced by Atomic Energy of Canada Limited and a French company. According to the level of technology at that time, it was a relatively complex system, including hardware and software. Due to defects in the software design, the instrument caused several medical incidents between June 1985 and January 1987 in which patients were exposed to excessive radiation, resulting in burns and even death. An investigation later found that the entire software system did not have strict quality assurance and had not been adequately tested. System security analysis only considers the hardware of the system, and does not take into account the hidden dangers caused by computer failures (including software problems).

What warnings should software quality assurance and software test workers get from this case?

2. Please analyze the faults in the following code.

```
public class getScoreAverage
{ public float getAverage( int [] scores )
  { if (scores==null || scores.length==0)
    {    throw new NullPointerException();
    }
      float sum = 0.0F;
      int j=scores.length;
    for (int i=1; i<j; i++)
    { sum += scores[i];
    }
    return sum/j;
    }
}
```

3. There are program segment as follows.

```
public int get_max(int x, int y, int z){
      int max;
      if(x>=y)
      {    max = x; }
      else
      {    max = y; }
      if( z>=x )
```

```
{     max = z;  }
return max;      }
```

(1) Try to analyze logical error in the program segment.

(2) Design a test data, so that the execution of the test will execute the defect code but will not trigger the error.

(3) Design a test data so that the execution of the test will execute the defect code and trigger the error, but will not cause failure.

(4) Design a test data so that the execution of the test will execute the defect code, trigger the error, and cause failure.

Chapter 2   Static white-box test

2.1   Overview of static white-box test

Static white-box test refers to the process of analyzing and checking the program code without executing the program, so as to find out problems and defects in source code, as shown in Figure 2-1.

```
public class gys
{ public int getGYS(int x,int y)
    { int Q=x;
      int R=y;
      while(Q!=
        { if (Q>R)
            Q=Q-R;
          else R=R-Q;
        }
        return Q;
    }
  public static void main(String[] args)
  { gys g = new gys();
    System.out.println(g.getGYS(63, 14));
  }
}
```

Figure 2-1   Static white-box test

The most fundamental check of a program is to detect syntax errors in the source code. This type of check can be performed by the compiler, which analyzes the program's syntax line by line, identifies errors, and reports them. However, there are many bugs or issues that are not grammatical errors and cannot be identified by the compiler. Therefore, developers or testers must use manual or automated methods to scrutinize and analyze the source code to find out these bugs or issues (as shown in Figure 2-2).

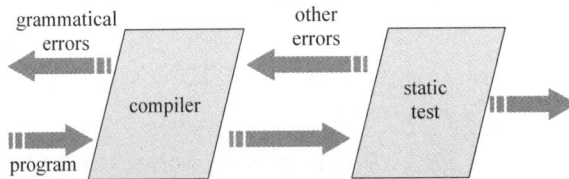

Figure 2-2   Static white-box testing does not aim at grammatical errors

Static white-box test identifies problems through examination and analysis of logic, structure, procedures, interfaces, coding specifications etc., within the program code. It aims to uncover flaws and suspicious elements such as mismatched parameters, improper loop or branch nesting, disallowed recursion, unused variables，references to null pointers, and suspicious calculations etc. The results of static white-box test can be utilized for further error checking and provide guidance for test case selection. Common types of static test include code review, static structure analysis, program flow analysis, static quality measurement etc.

2.2   Code review

2.2.1   Code check   代码检查

Code check involves directly checking the source code to ensure that it meets relevant specifications. In the past, this process took various forms such as desktop checks, code reviews, and program walk-through. With the development of technology, code checking is usually done automatically by test tool.

The main contents of the code inspection are as follows.

(1) Ensuring consistency between the code and its design.

(2) Verifying the correctness of logical expressions within the code.

(3) Assessing the readability of the code.

(4) Enforcing standardization and consistency in code style and format.

(5) Checking for compliance with coding rules and specifications.

(6) Evaluating the rationality of the code structure.

(7) Identifying unsafe, unclear, or ambiguous sections within the program.

Unlike dynamic test which can only detect external symptoms of errors, code check can precisely locate errors within the source code once they are identified. This ability ultimately reduces error-fixing costs. Additionally, during a thorough review of the source code, batches of errors may be uncovered, such as similar errors scattering multiple locations. In contrast, dynamic test typically can only identifies and reports error one by one.

2.2.2　Coding rules and programming specifications

In the process of programming, through code analysis and lessons learned, it is evident that adherence to certain rules can significantly reduce the likelihood of program problems. Conversely, failure to follow these rules may lead to program errors or other issues. These rules or guidelines are often referred to as coding rules and programming specifications. We can check that the code under test complies with these. In this way, the errors and omissions that are prone to occur in the process of software coding can be avoided as much as possible, and the quality of software products can be improved.

As software projects continue to grow in scale, it often becomes necessary for multiple individuals to collaborate on a single project in order to complete complex and large-scale software systems. In such cases, ensuring uniformity in program style and standardizing code writing becomes a challenge. This is where programming specifications come into play. Programming specifications require all programmers involved in a project to adhere to unified styles, formats, and specifications when writing program code. This approach helps us unify the code style and enhance readability of the code. Recognizing the importance of coding rules and programming specifications, many large software development companies have established their own programming standards. For example, Google has proposed coding standards for various languages (including Java, C++, Objective-C), while Alibaba Group has developed both Chinese and English versions of its Java program development manual along with providing a Java development plug-in module designed to help developers automatically detect any deviations from their coding standards.

Coding rules and programming specifications are categorized into various types. Some are recommended to be followed or used as a reference, while others are mandatory for compliance. Some may be established and implemented by a certain software development organization, and some may be generally accepted. Some may pertain to specific programming languages, and some may be language-agnostic. Additionally, some may be verified manually, while others can be analyzed and measured automatically.

1. Examples of common Java coding rule violations

Table 2-1 provides examples of common Java coding rule violations.

常见 Java 编码规
则违背示例

Table 2-1   Examples of common Java coding rule violations

Categories	Rules	Instructions	Code sample
input validation and representation	cross-site scripting: DOM	sending unauthenticated data to a Web browser causes the browser to execute malicious code	`insert = $(nNewNode);`
	cross-site scripting: persistent	sending unauthenticated data to a Web browser causes the browser to execute malicious code	`<button class="btn btn-mini btn-danger" type="button" onclick="dormBuildDelete(${dormBuild.dormBuildId})">Delete</button></td>`
	cross-site scripting: reflected	sending unauthenticated data to a Web browser causes the browser to execute malicious code	`<td><input type="text" id="dormBuildName" name="dormBuildName" value="${dormBuild.dormBuildName }" style="margin-top:5px;height:30px;" /></td>`
	SQL injection	constructing a dynamic SQL statement unreliable data source input allows attacker s to modify the meaning of statements or execute arbitrary SQL statements	`PreparedStatement pstmt = con.prepareStatement(sb.toString().replaceFirst("and", "where"));`
	header manipulation: cookies	contains unverified data, which may generate cookie manipulation attack, and lead to others HTTP responds header manipulation attacks	`Cookie user = new Cookie("dormuser", userName+"-"+password+"-"+userType+"-"+"yes");`
security features	password management: password in HTML form	filling a password filed in an HTML form can compromise system security	`<td><input type="password" id="password" name="password" value = "${dormManager.password}" style="margin-top:5px;height:30px;" /></td>`
	privacy violation	improper handling of confidential information, such as customer passwords or social security numbers, can compromise users' privacy	`<td>${dormManager.id}</td>`
environment	password management: password in configuration file	saving clear text passwords in configuration files may compromise system security.	`dbUserName=sa dbPassword=123456`

57

2. Common programming specifications

(1) Annotation.

① Annotations should be simple, clear and concise, with accurate meaning and prevent ambiguity.

② Annotate where necessary and in moderate amounts.

③ Modify the code at the same time as you modify the corresponding comments to ensure that the comments are consistent with the code.

④ The principle of proximity of comments, that is, keep the comment adjacent to its corresponding code, and should be placed above or with the code.

⑤ Global variables should have more detailed comments, including the description of its function, value range, which functions or processes access it, and precautions when accessing it.

⑥ In the head of each source file must have the necessary comment information, including: file name; Version number; Author; Date of generation; Module function description (such as function, main algorithm, the relationship between internal parts, the relationship between the file and other files, etc.); The main function or process list and the history of the file modification record.

⑦ Having the necessary comment information in front of each function or procedure, including: function or procedure name;Function description; Input, output and return value description; Calling relationship and called relationship description, etc.

(2) Naming.

① There should be uniform rules for naming.

② Avoid names that are difficult to understand.

③ Shorter words can be shortened by dropping the "vowel".

④ Longer words may be abbreviated by the first few characters of the word.

⑤ Use underscores to segment names that require more than one word.

(3) Variables.

① Get rid of unnecessary public variables.

② Construct a public variable that only one module or function can modify or create, and the rest related modules or functions can only access, so as to prevent multiple different modules or functions from modifying or creating the same public variable phenomenon.

③ Carefully define and clarify the meaning, function, value range and the relationship between public variables.

④ Clarify the relationship between public variables and functions or procedures that operate on this public variable, such as access, modification, and creation.

⑤ When passing data to public variables, be very careful to prevent assigning unreasonable values or crossing boundaries.

⑥ Prevent local variables from having the same name as public variables.

⑦ Carefully design the layout and order of the elements in the structure so that the structure is easy to understand, saves space, and reduces the phenomenon of causing misuse.

⑧ The design of the structure should try to consider forward compatibility and future version upgrades, and keep room for some possible future applications (such as reserving some space)

⑨ Pay attention to the specific programming language and the principles and details of how the compiler handles different data types.

⑩ Do not use uninitialized variables. You should initialize the variable at the same time you declare it.

⑪ When programming, be careful about casting data types.

(4) Function，Procedure.

① Try to limit the size of a single function to 200 lines.

② It is best if a function does only one thing.

③ Write functions for simple but commonly used functions.

④ Try not to write functions that depend on the internal implementation of other functions.

⑤ Minimize the parameters of the function to reduce the probability of errors when the function is called.

⑥ Use comments to detail the role of each parameter, the range of values, and the relationship between the parameters.

⑦ Check the validity of all parameters entered into the function.

⑧ The validity of all non-parametric inputs to the function, such as data files, public variables, etc., should be checked.

⑨ The function name should accurately describe the function's function.

⑩ The return value of the function should be clear, especially the meaning of the error return value should be accurate.

⑪ Clarify what the function does, the code should be able to accurately (not approximately) implement what it should do.

⑫ Reduce recursive calls to functions themselves or between functions.

⑬ When writing re-entrant functions, if global variables are used, they should be protected by turning off interrupts, semaphores (i.e., P, V operations)

(5) Code Testability.

① Funnel design，common logic normalization.

② Reduce the module coupling.

③ Program for interfaces, using functional interfaces to isolate external dependencies.

④ Before writing code, methods and means of program debugging and test should be designed in advance, and various debugging means and corresponding test code, such as test scripts, output statements, etc.

(6) Program Efficiency.

① Always pay attention to the efficiency of the code when programming. Especially code that needs to be executed repeatedly and concurrently.

② The code efficiency should be improved under the premise of ensuring the correctness, stability, readability and testability of the software system, instead of blindly pursuing code efficiency.

③ To carefully construct or directly use assembly language to write frequent calls or high performance requirements of the function.

④ Improve the efficiency of data structure division and organization of the system, and optimize the program algorithm.

⑤ In multiple loops, the busiest loop should be placed in the innermost layer.

⑥ Minimize the level of loop nesting.

⑦ Try to substitute multiplication or other methods for division, especially division in floating-point operations.

2.3　Static structure analysis　　　　　　　　静态结构分析

A piece of software typically consists of multiple components, each with a specific organizational structure and correlation. In static structure analysis, test tools are used to analyze the control logic, data structure, module interface, and calling relationships within the program code. This process involves generating various diagrams and tables such as control flow graphs, calling relationship diagrams, module organization charts, reference tables, and equivalence tables to clearly present the organizational structure and internal relations of the software. These visual tools facilitate both macro-level understanding and micro-level analysis of the program.

With the help of statistical charts, we can carry out control flow analysis, data flow analysis, interface analysis, expression analysis, etc. WE can find problems or unreasonable places in the program, and with further inspection we can confirm whether there are defects or errors in the software. Static structure analysis usually uses the following methods.

(1) Generate a variety of tables to analyze program code.

① Label cross-reference tables.

② Variable cross-reference table.

③ Subroutines (macros, functions) reference tables.

④ Equivalence table.

⑤ Constant table.

(2) Check whether the program has problems by analyzing various relationship diagrams and control flow graphs.

① Control flow graph: a graph composed of many nodes and edges connecting nodes, where each node represents one or more words.

Sentence, the edge represents the control flow direction, which can intuitively reflect the internal structure of a function.

② Function call relationship diagram: list all functions, use the line to indicate the call relationship, through the application between the call relationship between the functions to show the structure of the system.

③ File or page call relationship diagram.

④ Module structure diagram.

(3) Other common mistakes analysis. Analyze whether there are certain types of problems, bugs, or "dangerous" constructs in the program.

① Data type and unit analysis.

② Reference analysis.

③ Expression analysis.

④ Interface analysis.

2.4 Program flow analysis

程序流分析

A program must meet certain basic requirements in order to function normally and do not leave hidden dangers. The following is an analysis of the program's control flow and data flow.

1. Control flow analysis

控制流分析

From a control flow perspective, a program should adhere to the following basic requirements.

(1) It cannot jump to function，method，page，etc., that do not exist.

If jump to a function, method, page, etc., that does not exist, program execution will unexpectedly stop.

(2) There should be no statements that cannot be reached from the program entry.

If there are statements that cannot be reached from the program entry, it means that these statements will not be executed at all, and their corresponding functions cannot be invoked.

(3) There should be no statement，function，method，interface，etc., that cannot be exited from execution.

If there is statement, function, method that cannot be exited, it means that the program may be trapped in an endless loop. If there is an interface that cannot exit, it means that once a user enters the interface, he can not exit. For example, there is no exit button in the main interface of an APP, that will cause the user very inconvenient.

2. Data flow analysis

数据流分析

Data flow analysis is the process of analyzing the definition, reference, and dependencies between the data in a program. When a statement can change the value of variable V when executed, variable V is considered to be defined by the statement. When the value of variable V is used in a statement, the variable V is considered to be referenced by the statement.

The statement $X:=Y+Z$ defines the variable X and references Y, Z.

The statement if $Y > Z$ then... , references the variables Y and Z.

The statement READ *X* refers to the variable *X*.

The statement WRITE *X* defines the variable *X*.

In general, variables should be defined before they are used. If you will not use it, do not define it.

3. Examples   示例

Some compilers have the function of program control flow analysis and data flow analysis, and can give hints. For example, write the code shown in Figure 2-3 in Eclipse.

```java
📄 *GCD.java ⊠
 1  package gcd;
 2  public class GCD {
 3      public int getGCD(int x,int y){
 4          int Var1=0;
 5          if(x<1||x>100)
 6          {   System.out.println("The input data is out of range!");
 7              return -1;  }
 8          if(y<1||y>100)
 9          {   System.out.println("The input data is out of range!");
10          return -1;      }
11          int max,min,result = 1;
12          if(x>=y)
13          {   max = x;
14              min = y;        }
15          else
16          {   max = y;
17              min = x;        }
18          for(int n=1;n<=min;n++)
19          {   if(min%n==0&&max%n==0)
20              {       if(n>result)
21                      result = n;         }       }
22          System.out.println("common divisor:"+result);
23          return result;  }
24
25      public static void main(String[] args) {
26          GCD g = new GCD();
27          g.getGCD(5, 15);
28          g.getGBS(6, 8);
29          Var2 = 0;
30          }
31      }
```

Figure 2-3   Control flow and data flow analysis in Eclipse

Mouse over the exclamation mark 🔔 at the beginning of line 4, Eclipse will give prompt as follow.

The value of the local variable Var1 is not used

It is a reminder that the defined variable Var1 is not being used .

Mouse over the sign 🔳 at the beginning of line 28, Eclipse will give prompt as follow.

> The method getGBS(int, int) is undefined for the type GCD

It reminds that the method to be called getGBS() is not defined .

Mouse over the sign ▨ at the beginning of line29, eclipse will give prompt as follow.

> Var2 cannot be resolved to a variable

It reminds that the variable Var2 is not defined.

2.5   Static quality measurement   静态质量度量

Software quality measurement involves the assessment of each quality characteristic of software based on a specific quality model, and can give out a quality measurement result. With advancements in technology, automated test tools are commonly utilized to evaluate program code quality and generate a comprehensive measurement report. For instance, Figure 2-4 illustrates a sample quality report obtained from Logiscope, a static white-box test tool.

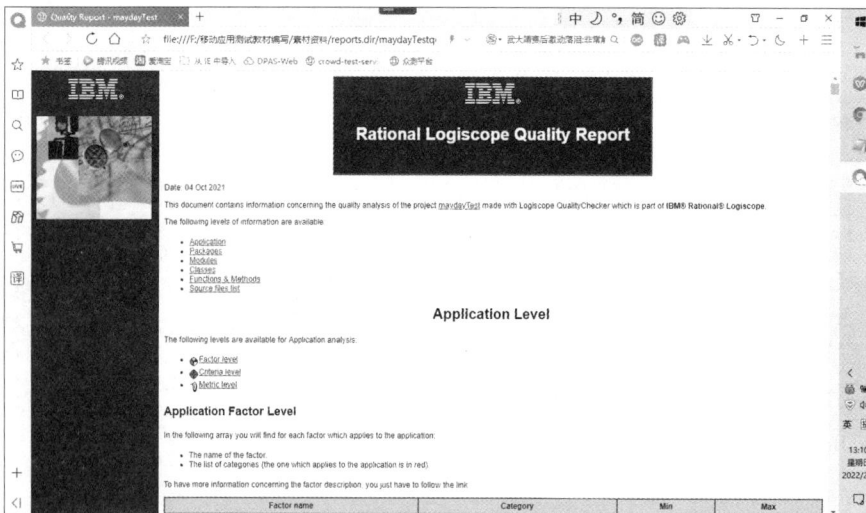

Figure 2-4   Logiscope quality report

2.6   Static white-box test example

静态白盒
测试实例

Early static white-box test was conducted by individuals. However, with the advancement of technology, now static white-box test is primarily carried out using automated tools. The fundamental principle involves constructing an analysis system that resembles a high-level compilation system to inspect and analyze the code of the program being tested. During the analysis process, the automated tool scans and analyzes the code to identify any deviations from coding rules and programming specifications, providing error or warning information as necessary. Some tools are also capable of generating system call diagrams and evaluating code quality based on a predefined software quality model. Currently, there are various static test (or static analysis) tools in use within the industry, and some are open source.

The following is an example of static white-box test using the SDAP tool software. The working interface of SDAP is shown in figure 2-5.

Figure 2-5   The working interface of SDAP

2.6.1   Test procedure

The basic test process is as follows.

(1) Create the test item in the SDAP Test Tool software.

(2) Perform the test project configuration, including the code development language under test, etc.

(3) Upload the source code compression package.

(4) Perform test .

(5) View or export the test report.

2.6.2　Test report

The test report generated by the SDAP test tool contains both summary information and specific content. The length of the test report is determined by the quantity and quality of the code tested, ranging from just a few pages to potentially thousands of pages.

1.Test report home page

Figure 2-6 shows the home page of the SDAP test report.

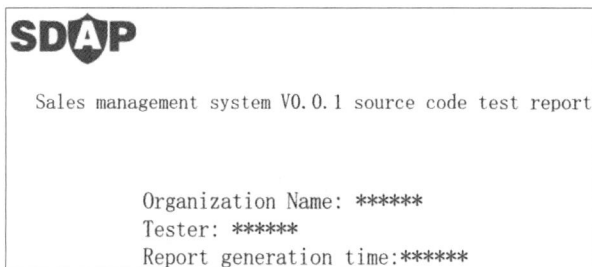

```
SDAP

Sales management system VO.0.1 source code test report



    Organization Name: *****
    Tester: *****
    Report generation time:*****
```

Figure 2-6　Home page of SDAP test report

2. Summary information

Figure 2-7 shows an example of SDAP test report summary information.

一、本次检测概况

项目名称：药品信息管理	版本：0.1.0
检测人：*****	检测时间：2021-09-23 03:35:28
检测耗时：7分钟	语言类型：Java
检测文件数：8	缺陷数：16
代码行数：1,672	缺陷密度：2.392 KLOC
严重等级的缺陷数：0	高等级的缺陷数：0
中等级的缺陷数：0	低等级的缺陷数：16

缺陷漏洞汇总表					
序号	所属领域	缺陷类型	缺陷级别	整改优先级	缺陷数
1	API Abuse	Unchecked Return Value	低	低	4
2	Encapsulation	Poor Logging Practice: Use of a System Output Stream	低	低	4

Figure 2-7　Sample SDAP test report summary information

3. Details

Figure 2-8 shows an example of the SDAP test report details.

二、检测结果详解

| 整改优先级：低 (16) |
| Unchecked Return Value（缺陷等级：低 [4]） |

缺陷描述：Java程序员常常会误解包含在许多java.io类中的read() 及相关方法。在 Java结果中，将大部分错误和异常事件都作为异常抛出。（这是 Java 相对于 C 语言等编程语言的优势：各种异常更加便于程序员考虑哪里出现了问题。）但是，如果只有少量的数据可用，stream和reader 类并不认为这是异常的情况。这些类只是将这些少量的数据添加到返回值缓冲区，并且将返回值设置为读取的字节或字符数。所以，并不能保证返回的数据量一定等于请求的数据量。

这样，程序员就需要检查read() 和其他 IO方法的返回值，以确保接收到期望的数据量。

在这种情况下，中第 行的 的值未进行验证。

示例：下列代码会在一组用户中进行循环，读取每个用户的私人数据文件。程序员假设这些文件总是正好 1000 字节，从而忽略了检查 read() 的返回值。如果攻击者能够创建一个较小的文件，程序就会重复利用前一个用户的剩余数据，并对这些数据进行处理，就像这些数据属于攻击者一样。

Figure 2-8　Example of the SDAP test report details

Exercise 2

Ⅰ. Single choice questions.

1. The naming requirements in the program do not include(　　).

　　A. There should be uniform rules for naming

　　B. Avoid names that are difficult to understand

　　C. Keep the name as short as possible

　　D. It is advisable to abbreviate the first few characters of a longer word

2. Common coding rule violations do not include(　　).

　　A. Storing clear-text passwords in a configuration file, which may compromise system security

　　B. Excessive comments may affect the execution efficiency of the program

C. Build dynamic SQL statements with input from unreliable data sources, allowing attackers to modify the meaning of statements or execute arbitrary SQL commands

D. Sending unauthenticated data to a Web browser can cause the browser to execute malicious code

3. The problems that static white-box test can find do not include( ).

A. Incorrect intermediate results

B. Mismatched parameters

C. Improper loop nesting and branch nesting

D. References to null pointers

Ⅱ. Fill in the blanks.

1. _____ analysis is the process of analyzing the definition, use, and dependencies between data in a program.

2. SDAP is a _____ white-box test tool.

3. _____ and _____ should be followed in programming. In this way, you can avoid as many errors and omissions as possible, and improve the quality of the software.

4. There is the necessary _____ information in the header of each source code file, including: file name, version number, author, date of generation, module function description, etc.

Ⅲ. True or false questions.

1. There should be no statements, functions, methods, interfaces, etc., that cannot exit execution.( )

2. The test report format of static white-box test is fixed, which has nothing to do with the quantity and quality of the tested code.( )

3. The software quality measurement of static white-box test test tool is to measure the quality characteristics of the software according to the requirements of a certain software, and give the measurement results.( )

Ⅳ. Comprehensive question.

1. Please point out the problems in the following program segment.

```java
import java.io.*;
public class Test {
  public boolean copy(InputStream is, OutputStream os) throws IOException {
    int count = 0;
    byte[] buffer = new byte[1024];
    while ((count = is.read(buffer)) >= 0)
      os.write(buffer, 0, count);
    return true;
```

```
    }
  public void copy(String[] a, String[] b, String ending) {
    int index;
    String temp = null;
    System.out.println(temp.length());
    int length = a.length;
    for (index = 0; index < a.length; index++) {
      if (true) {
        if (temp == ending)
          break;
        b[index] = temp;
} } }
  public void readFile(File file) {
    InputStream is = null;
    OutputStream os = null;
    try {
      is = new BufferedInputStream(new FileInputStream(file));
      os = new ByteArrayOutputStream();
      copy(is, os);
      is.close();
      os.close();
    } catch (IOException e) {
      e.printStackTrace();
    } finally {
} } }
```

2. The function of Fun1 is that convert the score of the hundred-mark to the score of the five-level scoring. Please analyze and point out the errors in the program, and then write out the complete and correct program.

```
void Fun1(int score)
{   if ( score >= 90 ) printf("Excellent\n");
    if ( ( score < 80 ) && ( score >= 70 ) ) printf("Medium\n");
    if ( (score <70) && ( score > 60 )) printf("Pass\n");
    if (score < 60) printf("Flunk\n");
 }
void main()
 { char number;
     scanf("%c", &number);
     Fun1(number);
 }
```

Chapter 3   Dynamic white-box test design

To identify software problems through dynamic test, the likelihood is low as indicated by the PIE model of software test. Test design is essential for enhancing the efficiency of test by minimizing unnecessary test cases. The fundamental methods for dynamic white-box test design include logical coverage, basic path coverage, loop test, mutation test, symbolic execution, and so on.

3.1   Logical coverage

3.1.1   Introduction

Logical coverage is one of the main dynamic test design methods in white-box test, which is a test technique based on the internal logical structure of the program. It means that all logical units, logical branches, and logical values used as test criterion should be executed. This approach requires the tester to have a clear understanding of the logical structure of the program. The criteria for logical coverage are statement coverage, decision coverage (another name is branch coverage), condition coverage, decision/condition coverage, and condition combination coverage. There is a program segment P1 as follows.

作为测试标准的逻辑单元、逻辑分支、逻辑取值都需要被执行到

```
If ( x>0 OR y>0 ) then a = 10
If ( x<10 AND y<10 ) then b = 0
```

The initial values of variables a, b have been defined already and are both -1. The program flow chart for the P1 is shown in Figure 3-1.

Next, we will see how to realize statement coverage, decision coverage, condition coverage, decision/condition coverage, and condition combination coverage.

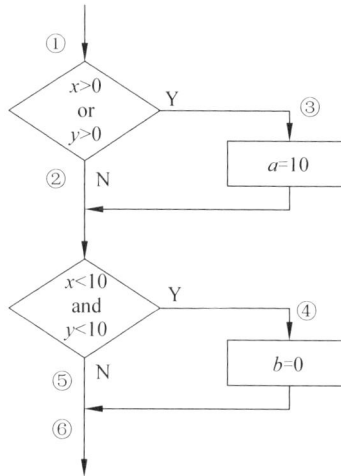

Figure 3-1   Program flow chart of P1

3.1.2   Statement coverage

语句覆盖

Statement coverage requires that test cases need to be designed so that each executable statement in the program can be executed at least once. For the program segment P1 whose flow chart is shown in Figure 3-1, the locations in the program needs to be covered or executed are ① ③ ④ ⑥, and the locations ② ⑤ do not need to be covered, since there are no statements. Two test cases can be designed to implement statement coverage. There all cover locations ① ⑥. In addition, one test case cover locations ③, where the first IF structure has result of "true", and the other test case cover locations ④, where the second IF structure has result of "true".

程序中的每个可执行语句至少都能被执行一次

Such as:

case1: $x=1$, $y=1$ cover ① ⑥ ③;

case2: $x=-1$, $y=-1$ cover ① ⑥ ④.

They can achieve the statement coverage requirements. From the perspective of saving test costs, we can optimize the design of test cases. In fact, only one test case is needed to cover ①③④⑥.

Such as:

case3: $x=8$, $y=8$

Their execution path is shown in Figure 3-2.

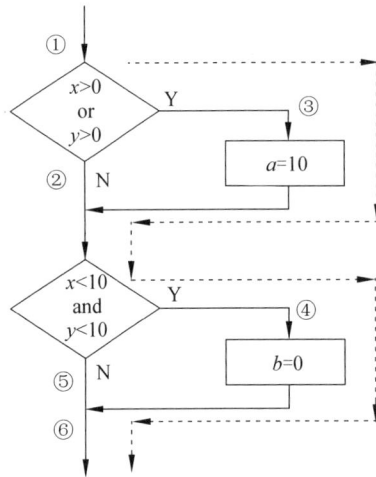

Figure 3-2　Statement coverage execution path

For a software with a certain scale, it may be not easy to achieve 100% statement coverage. For example, some codes are used for error handling or for coping with certain special situation, if such error or special situation does not occur, these codes will not be executed. At this time, in order to improve the statement coverage, it is necessary to design test cases specifically.

Statement coverage is a weak coverage criterion. As you can see from Figure 3-2, in both judgment statements, only one branch is executed, while the other branch is not executed at all. Statement coverage tests every executable statement in the program, which seems to be able to test the program comprehensively, but in fact, it is not a very sufficient coverage standard for test, and sometimes some obvious errors can not be found by statement coverage test.

If in the program section P1, the logical operation symbols of two judgment statements are written wrong due to negligence, the "OR" in the first judgment statement is mistakenly written as "AND", and the "AND" in the second judgment statement is mistakenly written as "OR", and the test case case3 is used for test, the execution path is still ①③④⑥, as shown in Figure 3-3, and the test result is still correct. The test failed to find errors in the program.

The advantage of statement coverage is that it is relatively simple to analyze and apply. The disadvantage is that it is insensitive to the control structure, the coverage of the program execution logic is inadequate, and it may fails to find errors. The statement coverage is calculated as follows.

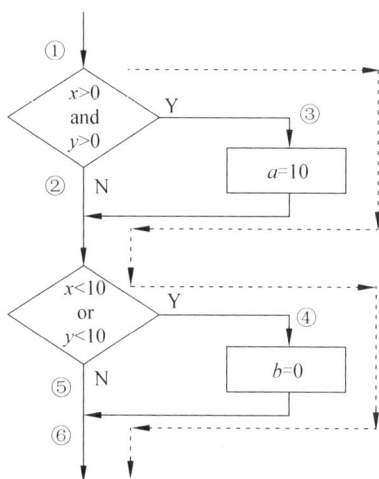

Figure 3-3   Statement coverage test fails to find errors

Statement coverage = number of executable statements tested / total number of executable statements × 100%.

3.1.3   Decision coverage

判定覆盖

A slightly stronger coverage criterion than statement coverage is decision coverage. Decision coverage is to design some test cases and run the program under test, so that the true value result and false value result of each judgment appear at least once in the program. Decision coverage is also referred to as branch coverage, because if the judgment is true, the true branch will be executed, and if the judgment is false, the false branch will be executed. The true and false consequences of each judgment occur at least once which is equivalent to the true and false branches of each judgment occurring at least once.

使得程序中每个判断的真值结果和假值结果都至少出现一次

Take program segment P1 as an example, according to the requirement of decision coverage, the program needs to be executed through the location ①② ③④⑤⑥ in the flow chart. There are two IF statements in P1. Since each judgment has both true and false results, two test cases are required at least. The two IF statements in P1 are concatenated, not nested, so two test cases are indeed enough if properly designed. The following two test cases can meet the decision coverage requirements.

case4: $x= 20$, $y= 20$,   cover ①③⑤⑥;

case5: $x= -2$, $y= -2$,   cover ①②④⑥.

You can see Table 3-1 for coverage details.

Table 3-1   Decision coverage table

Test case number	x	y	1st judgment expression $x>0$ or $y>0$	2nd judgment expression $x<10$ and $y<10$
case4	20	20	Y	N
case5	−2	−2	N	Y

If a test achieves decision coverage, it indicates that all branches in the program flow chart will be tested, and all statements on each branch will be tested as well. Therefore, as long as decision coverage is met, statement coverage will also be met. In decision coverage, when there are multiple conditions in a judgment expression, only the final result of the judgment is focused on rather than the result of each individual condition. As a result, some conditions may always take only one value while the other value does not appear at all. At this point, even if the conditional expression is written incorrectly, the decision coverage test may not find it.

In other words, when the decision expression in the program is composed of several conditions, the decision coverage of the test for each condition is not sufficient, and it may not be able to find the possible errors in each condition. The calculation formula for determining coverage is as follows.

decision coverage = Number of judgment branches tested / Total number of judgment branches × 100%.

3.1.4   Condition coverage

条件覆盖

Condition coverage requires each condition in the judgment expression to obtain at least true value once and false value once. It should be noted that each condition obtains at least true value once and false value once is not equal to each judgment can obtain at least true value once and false value once. Condition coverage is not stronger than decision coverage, and they just have different focuses. There is no strict strong and weak relationship between them. For program segment P1, the following test cases can be designed to achieve condition coverage.

判断表达式中的每一个条件都要至少取得一次真值和一次假值

case6: $x = 20$，$y = -20$;

case7: $x = -2$, $y = 20$.

The specific coverage is shown in Table 3-2.

Table 3-2　Condition coverage test case table

Test case number	x	y	Condition x>0	Condition y>0	Condition x<10	Condition y<10
case6	20	−20	Y	N	N	Y
case7	−2	20	N	Y	Y	N

Attention please, case6 and case7 cover only the Y branch of the 1st IF statement, and only the N branch of the 2nd IF statement, and therefore do not meet the decision coverage. The formula for calculating condition coverage is as follow.

Condition coverage = Number of condition values tested / Total number of condition values × 100%.

3.1.5　Conditional/decision coverage

条件/判定覆盖

Condition coverage is not stronger than decision coverage. Their focus are different. Sometimes condition coverage and decision coverage are used together, which is called conditional/decision coverage. It means designing enough test cases so that the true/false value of each condition in the judgement expression occurs at least once, and the true/false value of each judgement expression also occurs at least once.

使得判定表达式中每个条件的真/假取值至少都出现一次，并且每个判定表达式自身的真/假取值也都要至少出现一次

For program segment P1, the test cases case4 and case5, that we design when doing the decision coverage, meet condition/decision coverage at the same time, because the true/false values of each condition appear once and the true/false results of each judgment also appear once. The specific coverage is shown in Table 3-3.

Table 3-3　Case4, case5 meet condition coverage

Test case number	x	y	Condition x>0	Condition y>0	Condition x<10	Condition y<10
case4	20	20	Y	Y	N	N
case5	−2	−2	N	N	Y	Y

There is program segment P2 as follows, which is used for triangle determination problem.

```
if ((a<b+c) && (b<a + c) && (c<a + b))
  is_Triangle = true ;
else
  is_Triangle = false ;
```

When testing this program segment, all four expressions (shown in Table 3-4) must have both true and false values if the condition/decision coverage is to be satisfied.

Table 3-4   Four expressions

Expression number	Expression
1	$a<b + c$
2	$b<a + c$
3	$c<a + b$
4	$(a<b+ c)$ && $(b<a +c)$ && $(c<a +b)$

The following test cases can be designed to satisfy the condition/decision coverage.

case8: $a=1, b=1, c=1$;

case9: $a=1, b=2, c=3$;

case10: $a=3, b=1, c=2$;

case11: $a=2, b=3, c=1$.

The specific coverage is shown in Table 3-5.

Table 3-5   Test cases that satisfy conditional/decision coverage

Test case number	a	b	c	Expression 1	Expression 2	Expression 3	Expression 4
case8	1	1	1	Y	Y	Y	Y
case9	1	2	3	Y	Y	N	N
case10	3	1	2	N	Y	Y	N
case11	2	3	1	Y	N	Y	N

The formula for calculating the condition/decision coverage is as follows.

Conditional/decision coverage = Number of conditional values and decision branches tested / (Total number of conditional values +Total number of decision branches) × 100%.

3.1.6 Condition combination coverage

条件组合覆盖

Condition combination coverage is also called multi-condition coverage, which means that enough test cases should be designed so that all various combinations of conditional values in each judgment occur at least once. Obviously test cases that satisfy condition combination coverage can also satisfy decision coverage, condition coverage, and condition/decision coverage.

每个判定中条件
取值的各种组合
都至少出现一次

For program segment P1, since there are two conditions in a judgment, there are four possible combinations of the two conditions, and at least four test cases are needed to achieve condition combination coverage. If they can be designed reasonably, cover the four combinations of conditions in the first decision and cover the four combinations of conditions in the second decision, four test cases will be enough to meet the condition combination coverage.

Such as:

case12: $x= 50$, $y= 50$;

case13: $x= -5$, $y= -5$;

case14: $x= 50$, $y= -5$;

case15: $x= -5$, $y= 50$.

The condition combination coverage for the 2 judgment expressions is shown in Table 3-6.

Table 3-6 Condition combination coverage

Test case number	x	y	1st decision		2nd decision	
			Condition $x>0$	Condition $y>0$	Condition $x<10$	Condition $y<10$
case12	50	50	Y	Y	N	N
case13	−5	−5	N	N	Y	Y
case14	50	−5	Y	N	N	Y
case15	−5	50	N	Y	Y	N

The four test cases above that satisfy condition combination coverage do not necessarily cover every possible execution path in the program. For example, path ①②⑤⑥ as shown in Figure 3-4 are not covered.

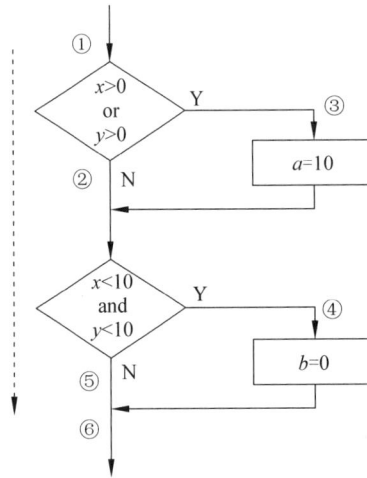

Figure 3-4   Condition combination coverage fails to cover some execution paths

The formula for calculating condition combination coverage is as follows.

Condition combination Coverage = Number of Conditional Value Combinations Tested/Total Number of Conditional Value Combinations × 100%.

If a judgment expression consists of 4 conditions, then 2^4 or 16 test cases need to be designed to implement condition combination coverage. If a judgment expression consists of 6 conditions, then 2^6 or 64 test cases need to be designed. The disadvantage of condition combination coverage is that when there are many conditions in a judgment statement, the number of condition combinations will be large, requiring many test cases. In terms of ease of test, when writing a program, the number of conditions in a decision expression should not be too large.

3.1.7   Summary of coverage criteria

覆盖标准小结

Coverage criteria describe the degree to which the software under test is tested. It is sometimes referred to as test data completeness criterion, which can be used to measure the adequacy of the test, can be used as one of the criteria for stopping the test, and is also the basis for selecting the test data. Test data sets that fulfill the same coverage criteria are considered equivalent. Different coverage criteria have different test adequacy. The test adequacy of statement coverage, decision coverage, condition coverage, condition/decision coverage,

不同的覆盖准则
具有不同的测试
充分性

and condition combination coverage has strong and weak relationship as shown in Figure 3-5.

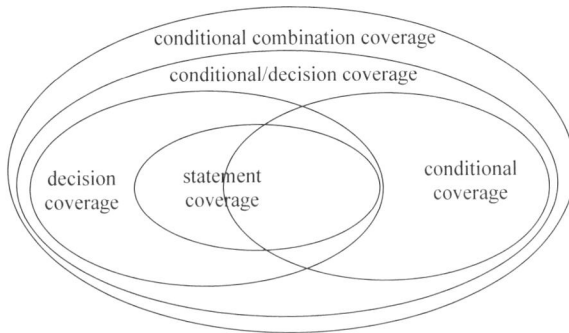

Figure 3-5   The relationships of logical coverage criteria

If coverage criterion A has higher adequacy than coverage criterion B, it implies that the set of test cases satisfying coverage criterion A will also satisfy coverage criterion B. The test adequacy of statement coverage, decision coverage, condition coverage, condition/decision coverage, and condition combination coverage has strong or weak relationship as shown in Figure 3-5. Decision coverage is higher than statement coverage. However, condition coverage does not necessarily surpass statement coverage (as illustrated in Figure 3-5).

1. The role of testing coverage criteria

测试覆盖准则的
作用

The role of test coverage criteria is reflected in the following aspects.

(1) The requirements and workload of software test can be defined quantitatively.

When testing a program, different test standards correspond to vary test requirements and workloads. For example, a small program may require 8 test cases for condition combination coverage, while only 2 test cases may be needed for condition coverage, as the former standard is more stringent than the latter.

(2) It can reflect the adequacy of the test.

By utilizing logical coverage standards and corresponding coverage statistics, the extent of test can be assessed. A higher coverage standard indicates a more thorough test. Additionally, a higher coverage rate signifies greater adequacy in test. For instance, within statement coverage tests, achieving 100%

coverage is considered more adequate than reaching 95%.

(3) It serves as the basis for selecting test data.

In software test, an abundance of test data needs to be designed or selected. The coverage standard provides guidance for this process. Different logical coverage criteria necessitate distinct test data.

(4) It can be used as a criterion to stop test.

Test can not continue indefinitely and excessive test is wasteful. There are various reasons to halt a test, and meeting specific logical coverage criteria is such a reason. For example, if a software test requires condition coverage, once this standard has been met, it signals completion of the task and allows for cessation of further test efforts.

(5) It has a significant impact on test results and the evaluation of software quality.

Test results are closely tied to the test standards. Different coverage criteria applied to the same software may yield different test results. A software program may pass the test of a certain coverage criteria, but may not necessarily pass another. According to the different coverage criteria passed by the test, different evaluation opinions can be given to the software quality.

2. Test coverage rate   测试覆盖率

In software test, it is essential to calculate coverage rate according to specific test coverage criteria. The purpose of this is as follows.

(1) Improve test efficiency.

Through statistical analysis of coverage, redundant or invalid test data can be identified and eliminated, so as to improve test efficiency. For example, when two test engineers collaborate on designing test cases, let's call them Mr. Zhang and Mr. Li, coverage statistics analysis might reveal that some test cases designed by Mr. Li do not contribute to improving coverage rate when merged with those designed by Mr. Zhang. These redundant cases should be removed in order to reduce unnecessary workload.

(2) Find more problems and improve product quality.

Coverage statistics enable a clear description of how extensively a program has been tested; this allows for identification of untested or inadequately tested areas within the software so that further test can uncover more issues and enhance product quality. For instance, through coverage analyzing, it might become apparent that module X has a 0% coverage rate，which means that the module was not tested at all, and module Y only has 30% coverage rate, which means insufficient. In such instances, continued test for modules X and Y would be necessary.

3.2 Basic path coverage

基本路径覆盖

In black-box test, conducting exhaustive test on all possible input data is not feasible. Similarly, in white-box test, performing exhaustive path test on a software of a certain scale is also impractical. Therefore, we can only select a portion of all potential execution paths for test, and basic path coverage serves as such a method. When conducting structural analysis in a program, especially for basic path coverage, it is essential to utilize the control flow graph.

3.2.1 Control flow graph

控制流图

Control flow graph, which is used to describe the control flow of a program, is an abstract expression of a process or program. A control flow graph is a directed graph, and formally expressed as follows.

G = (N, E, N_entry, N_exit).

N is the set of nodes, each statement in the program corresponds to a node in the graph, and sometimes a set of sequentially executed statements with no branching can be combined to be represented by a single node.

E is the set of edges.

E = {< n1，n2 > | n1, n2 ∈ N, and n2 is executed immediately after n1}.

N_entry and N_exit are the entry and exit nodes of the program, and G has only a unique entry node N_entry and a unique exit node N_exit.

Each node in G can have two direct successors at most. For a node v with two direct successors, its exit edge has the attribute "T" or "F" respectively. For any node n in G, there is a path from N_entry through n to N_exit.

In a control flow graph, nodes are used to represent operations, condition judgments, and convergence points, and arcs or control flow lines are used to represent the sequential order of execution. The symbols of the control flow graph corresponding to the basic control structure of the program are shown in

Figure 3-6.

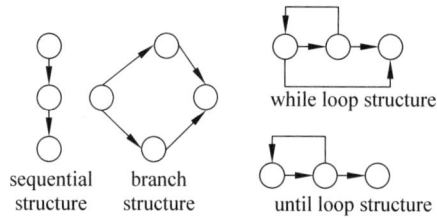

Figure 3-6　Control flow graph corresponding to the basic control structure

In the graphical notation shown in Figures 3-6, a circle is called a node of the control-flow graph, and it represents one or more statements without branch.

The directed arrow is known as an arc or control flow line, denoting the sequential relationship of execution. The control flow graph can be derived from a program or converted from a program flow chart.

The following two points need to be noted.

(1) When converting a program flow chart into a control flow graph, there should be a convergence node at the convergence of branches.

(2) If the condition expression in a judgment is a compound condition expression connected by logical operator, it needs to be decomposed into a series of nested judgments with only one condition.

如果判定中的条件表达式是由逻辑操作符连接的复合条件表达式，需要将其拆解为一系列只有单一条件的嵌套判断

Look at examples as below. In Figure 3-7, a partial program flow chart is

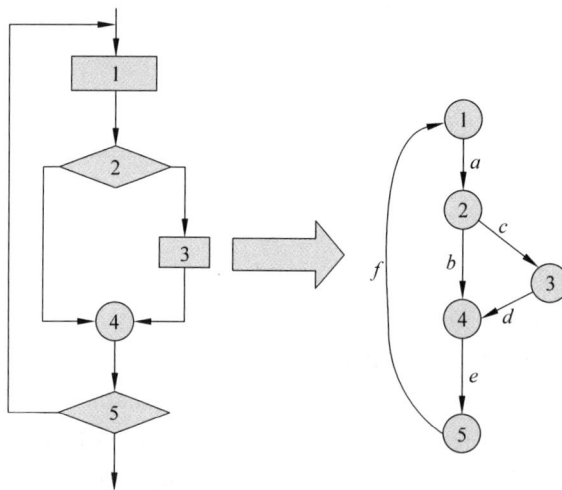

Figure 3-7　Convert branch convergence point to a node

shown on the left, and a control flow graph obtained by conversion from the left chart is shown on the right. In the left diagram there is no node at the position labeled ④, just a convergence point of the branch, but in the right graph the position labeled ④ is a node, which is converted from the convergence point of the branch.

In Figure 3-8, the left side is a program flow chart, and the right side is the control flow graph obtained by conversion from the left diagram.

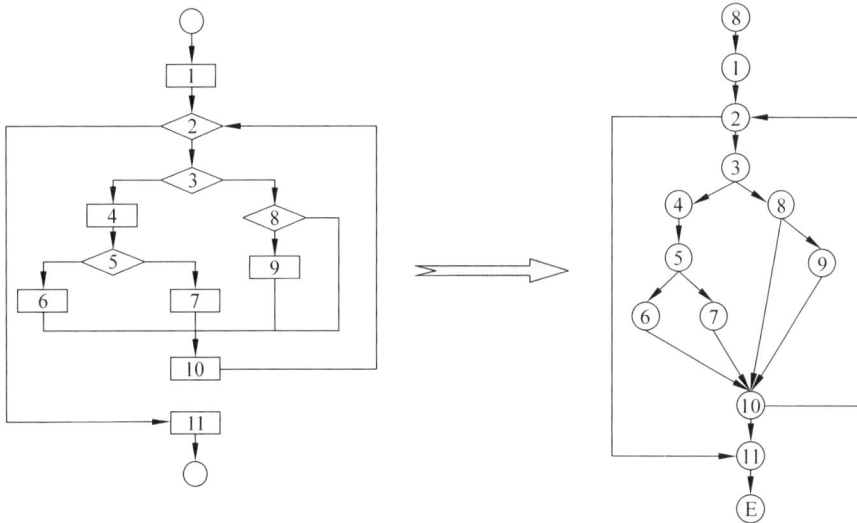

Figure 3-8   Program flow chart and the control flow graph obtained by conversion

In Figure 3-9, the left side shows a partial program flow chart with a multi-condition judgment box, and the right side shows a control flow graph obtained by taking apart the multi-condition judgment into two single condition judgments.

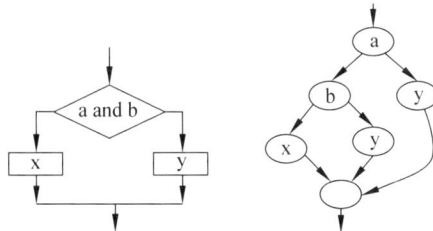

Figure 3-9   Take apart multi-condition judgment into multiple single condition judgments

3.2.2　Loop complexity　　　　　　　　　　　　　环路复杂度

How to measure the complexity of a program? Can the size of a program accurately reflect its complexity? Is a 1,000-line program necessarily more complex than a 100-line program? The answer is no. This is analogous to solving 100 addition and subtraction problems, which is not inherently more complex than solving a single binary integral problem. For instance, a program consisting of 1,000 lines of sequentially executed assignment and output statements does not inherently possess greater complexity than a 100-line sorting algorithm program. Measuring the complexity of a program solely in terms of its size is one-sided and inaccurate.

One method for measuring the complexity of a program is loop complexity. The more complex the control path is and the more loops there are in the program, the higher the loop complexity is. The loop complexity is used to quantitatively measure the logic complexity of the program. According to the control flow graph of the program, the loop complexity of the program can be calculated.

The loop complexity can be determined using three different methods based on the control flow graph.

(1) The loop complexity is the number of regions in the control flow graph. The area circled by edges and nodes is called region, and in addition the outside area which is not closed is also a region.

(2) Let E be the number of edges in the control flow graph and N be the number of nodes in the chart, then the loop complexity is $V(G) = E-N+2$.

(3) If we set P as the number of judgment nodes in the control flow graph, then the loop complexity $V(G) = P+1$.

For the same control flow graph, all three methods calculate the same result. Look at an example as below. In Figure 3-10, the flow chart of a program is shown on the left and the corresponding control flow graph on the right.

The loop complexity is calculated as follows using 3 methods respectively.

(1) The number of regions in the graph is 4, so the loop complexity $V(G) = 4$.

(2) The number of edges $E = 11$ and the number of nodes $N = 9$, the loop complexity $V(G) = E-N+2 = 4$.

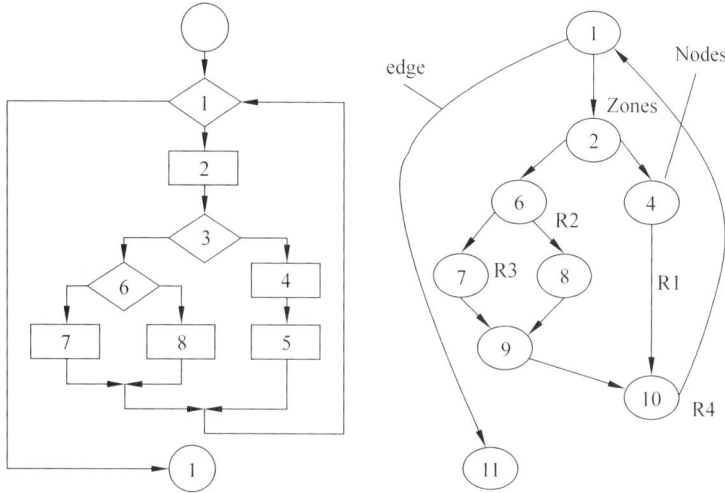

Figure 3-10   Program flow chart and corresponding control flow graph

(3) The number of judgment nodes $P = 3$ in the graph, the loop complexity $V(G) = 3+1 = 4$.

The three methods get the same results, and the loop complexity is 4.

3.2.3   Basic path coverage

基本路径覆盖

1. Path in program

程序中的路径

After abstracting a program as a directed graph, an ordered arrangement of the individual nodes passing from the entrance to the exit of the program is called a path, and a path expression can be used to represent such a path. A path expression can be a sequence of nodes or a sequence of arcs. For example, Figure 3-11 is a program control flow graph named CFG-A, and its possible pro-

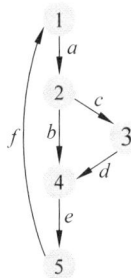

Figure 3-11   Program control flow graph CFG-A

gram execution paths are shown in Table 3-7.

Table 3-7   Possible program execution paths

Path number	Arc sequence representation	Node sequence representation
1	*acde*	1-2-3-4-5
2	*abe*	1-2-4-5
3	*abefabe*	1-2-4-5-1-2-4-5
4	*abefabefabe*	1-2-4-5-1-2-4-5-1-2-4-5
5	*abefacde*	1-2-4-5-1-2-3-4-5
… …	… …	… …

It should be noted that when there is a loop in the program, if the number of cycles executed by the program is different, then the corresponding execution path is different. In Table 3-7, paths 3 and 4 are examples of this. To enhance the expression ability, plus and exponential operation can be introduced into the path expression. Plus can express branching structure and exponential operation can express looping structure. Given a control flow graph as shown in Figure 3-12 named CFG-B, all of its possible paths can be expressed as $(ac + bd)e(fe)^n$, in which n is the number of loops.

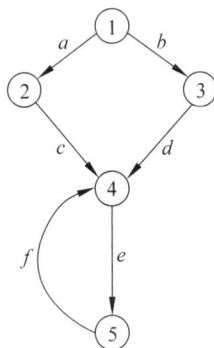

Figure 3-12   Program control flow graph CFG-B

2. The path exhaustion test is not feasible

路径穷举测试不可行

An IF statement results in two paths. The concatenation of two IF statements results in four paths. The number of possible paths for even a less complex program is a huge. If there are loops in program, the number of possible paths may be an astronomical figure. Figure 3-13 shows a program flow chart. If the upper limit of execution for each loop is 10 times, how many possible execution paths are there? The total number of possible execution paths L is calculated as follows.

$$L = 4^0 + 4^1 + \cdots + 4^{10} = 1,398,101$$

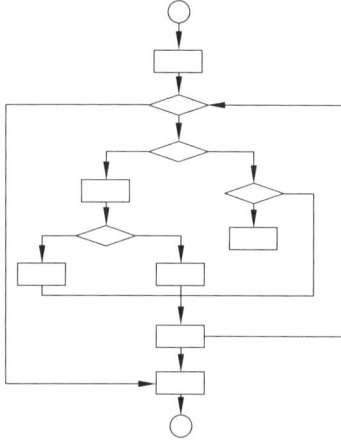

Figure 3-13   A program flow chart

Assuming that all possible paths in Figure 3-13 are executable, and a computer takes about 10 microseconds to execute this program once, 24 hours a day, 365 days a year, no downtime, the approximate time required to test all the paths is as follows.

$$1,398,101 \times 10/1,000,000 \approx 14 \text{ (s)}$$

If the upper limit of execution of each loop is 20 times, it will take about 4,072 hours, and the upper limit of 100 loops will take about $6.79 \times 1,047$ years. Paths exhaustive test is not feasible.

3. Basic path coverage   基本路径覆盖

Since it is difficult to test all paths in a program, we can only select part of the paths for test. Basic path test is a method that involves analyzing the loop complexity based on the program's control flow graph, identifying independent execution paths, also called basic paths, and designing test cases to cover these basic paths.

Basic path coverage ensures that all nodes in the program are covered, meaning that each executable statement will be executed at least once. Therefore, if basic path coverage is achieved, statement coverage will also be satisfied. The main steps of basic path coverage are as follows.

(1) Draw a control flow graph for the program.

(2) Calculate the loop complexity of the program.

(3) Determine the set of independent paths.

An independent path means that at least one path node is new and not included in other independent paths. The number of independent paths can be derived from the loop complexity of the program. It is equal to the loop complexity. This is a lower bound on the number of test cases necessary to ensure that each executable statement in the program is executed at least once.

After the number of independent paths of the program has been derived, based on the control flow graph, we can get each independent path. All independent paths form the set of independent paths, also known as the set of basic paths.

(4) Design test cases for each independent path.

Designing test cases ensures that each path in this set can be executed. Typically, one test case is designed for each independent path to guarantee its execution.

4. Example of basic path coverage

基本路径覆盖示例

A program segment IsLeap is provided as follows.

```java
public class Is_Leap {
    int leap = 0;
    int IsLeap(int year){
        if(year % 4 == 0){
            if(year % 100 == 0){
                if(year %400 == 0)
                    leap = 1;
                else    leap = 0;
            }else
                leap = 1;
        }else      leap = 0;
        return leap;        } }
```

For the program segment IsLeap, let the value range of year be 1000-9999, design test cases for the variable year to satisfy the basic path coverage.

(1) Draw a control flow graph corresponding to the program code, as shown in Figure 3-14.

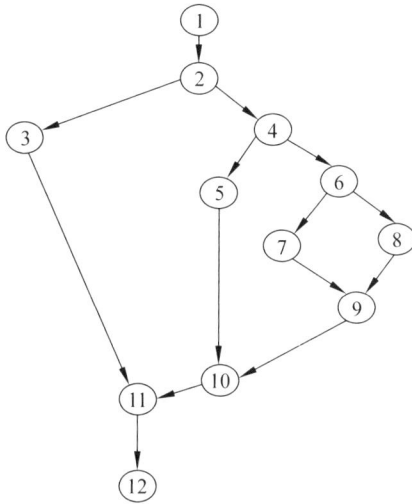

Figure 3-14   Control flow graph corresponding to the IsLeap program segment

(2) Compute the loop complexity $V(G)$.

$V(G) = E - N + 2 = 14 - 12 + 2 = 4$

$V(G)$ = number of judgment points + 1 = 3 + 1 = 4

$V(G)$ = number of regions = 4

(3) Determine the set of independent paths as follows.

① 1-2-3-11-12;

② 1-2-4-5-10-11-12;

③ 1-2-4-6-7-9-10-11-12;

④ 1-2-4-6-8-10-11-12.

(4) Design test cases.

Design test cases for each independent path as shown in Table 3-8.

Table 3-8   Set of test cases to satisfy the basic path coverage

Test case number	Test data	Expected execution result	Test path
1	year=1001	leap=0	1-2-3-11-12
2	year=1004	leap=1	1-2-4-5-10-11-12
3	year=1100	leap=0	1-2-4-6-7-9-10-11-12
4	year=2000	leap=1	1-2-4-6-8-10-11-12

3.3   Loop test

In the basic program structure, the loop is the most complex element. The expansion of the program's execution path is mainly caused by the loop structure. Different actual cycle numbers in a loop structure will result in different execution paths. Due to the uncertainty of the number of times the loop structure is executed during program execution, various situations may arise and errors are more likely to occur. Therefore, test should focus on the loop structure.

It is necessary to pay attention to and analyze the correctness of the loop structure in the program, and test the loops to verify that they work correctly in different situations, ensuring overall program correctness.

3.3.1   Basic loop structure test

There is a simple loop structure program code as follows.

```
int i=1, s=0, a=100;
while (i<=a)
{ s=s i.
  i=i 1; }
```

For such a basic loop structure, there are two common test methods, Z-path coverage test and loop boundary condition test.

1. Z-path coverage test

The Z-path coverage is a simplification of the loop mechanism. The simplification is to limit the number of loops. Regardless of the form of the loop, and regardless of the number of times the loop body may actually execute, only two cases that the loop condition is not satisfied and is satisfied only once are considered. Z-path coverage is equivalent to reducing the loop structure to a decision structure, as shown in Figure 3-15.

When testing a program, if Z-path coverage is used to limit the number of loops, the total number of execution paths of the program may not be too large, and then it is possible to achieve full coverage of all the paths of the simplified loop structure, which is what path enumeration is intended to do.

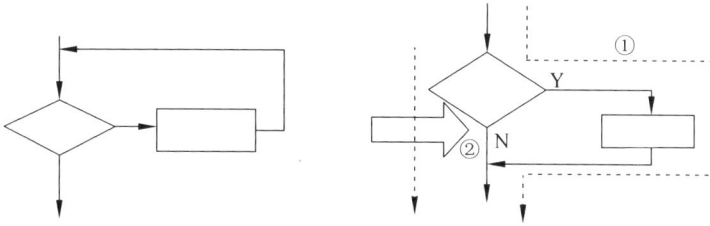

Figure 3-15　Z-path coverage reduces the loop structure to a decision structure

2. Loop boundary conditions test

循环边界条件测试

The second method of loop test is loop boundary condition test, which is equivalent to boundary value test of the loop count variable, typically covering seven boundary value points.

Let i be the actual number of loops and n be the maximum number of loops, then a loop boundary condition test should include the following test cases.

(1) Skip the loop body , $i=0$.

(2) Execute the loop body only once, $i=1$;

(3) Execute the loop body twice, $i=2$;

(4) Execute the loop body m $(2 < m < n-1)$ times, $i = m$;

(5) Execute the loop body $n-1$ times, $i = n-1$;

(6) Execute the loop body n times, $i = n$;

(7) Exceeding the maximum number of loops.

This is effectively equivalent to performing a 7-point boundary value test on the loop count variable, as shown in Figure 3-16.

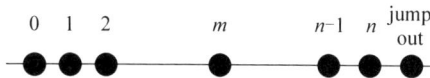

Figure 3-16　7-point boundary value test on the loop count variable

The following is an example of the application of cyclic boundary condition test. There is a program segment with a loop. We need to use cyclic boundary condition test method to test it. The program segment is as follows.

```
// Program under test
My_Sum { int j }
int i=1,s=0,a=100.
while (i<=j and i<=a)
```

```
{ s=s i.
  i=i 1; }
```

To do 7-point cyclic boundary condition test, test data designed is as follows.

case1: j=0 actual loop 0 times

case2: j=1 actual loop 1 time

case3: j=2 actual loop 2 times

case4: j=50 actual 50 cycles

case5: j=99 actual cycle 99 times

case6: j=100 actual cycle 100 times

case7: j=101 actual loop 100 times, and then i=101 exceeds the maximum number of loops.

3.3.2　Compound loop structure test　复合循环结构测试

Apart from the basic loop structure, compound loop structures may occur in the program. They are connected loops, nested loops, unstructured loops nesting.

These three compound loop structures are shown in Figure 3-17.

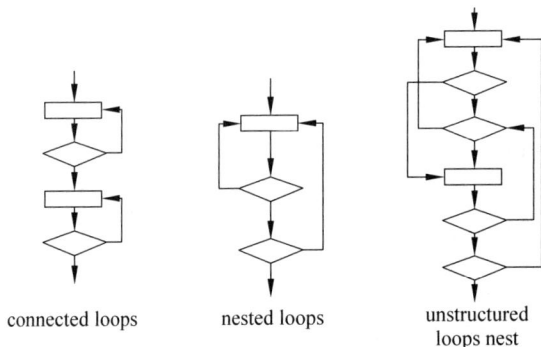

connected loops　　nested loops　　unstructured loops nest

Figure 3-17　Compound loop structure

1. Connected loops test　连接循环测试

Refers to two or more simple loops connected in series for sequential execution.

If the connected loops are independent from each other, it is sufficient to test each loop body according to the basic loop. If the final result of the loop variable of connected loop body 1 is the initial value of the loop variable of loop body 2, then the method for test nested loops can be used.

2. Nested loops test   嵌套循环测试

A loop structure which contains another loop structure.Nested loops are tested as follows.

(1) Starting from the innermost level, the loop variables in the other levels are set to their minimum values.

(2) Test the innermost loop body in the same way as for a simple loop, with the outer loop still taking the minimum value.

(3) Extend the loop body outward to test the next loop.

(4) All outer loop variables are minimized.

(5) The remaining inner nested loop bodies take typical values.

(6) Continue this step until all loop bodies have been tested.

3. Unstructured loops nesting test   非结构循环测试

Refers to a situation where exist direct jump from one loop body to another loop body. Test unstructured loops nesting can be a big headache, and it is best to redesign the structure of the loop body so that it becomes a nested loop or a connected loop.

3.4   Program mutation test   程序变异测试

In the test of software, we suppose that designing and executing a large number of test data without encountering any problems, and the execution results are correct. At this point, there are two possibilities to consider. The first is that the software is of high quality with minimal issues. The second is that the quality of our designed test data is poor, making it difficult to identify program errors, as illustrated in Figure 3-18.

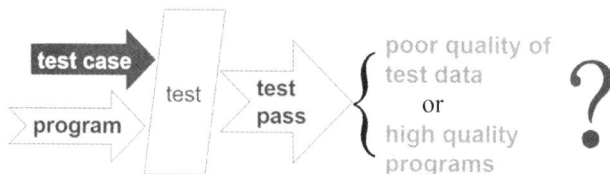

Figure 3-18   The software quality is high or test data quality is poor

To verify which possibility holds true, we can artificially modify the program according to certain rules to introduce errors. Subsequently, we run the program using the previous test data to determine if any execution errors occur. If errors do arise, it indicates that our test data is effective; however, if no errors manifest, then it suggests that our test data lacks sufficient quality as depicted in Figure 3-19. This process aids in understanding mutation test and its purpose. It should be noted that this example only scratches the surface of what mutation test entails; there are more comprehensive aspects to consider beyond this simple case.

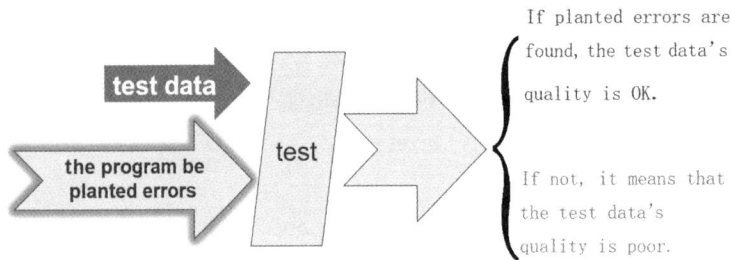

Figure 3-19   Schematic diagram of mutation test

3.4.1   Program mutation

程序变异

Program mutation is typically an operation that introduces slight changes to a program in order to test the validity of the test data by intentionally introducing errors. The most realistic and useful type of modification for checking the validity of test data and improving its quality is one that simulates common errors and omissions within the program. Test aims to identify errors and omissions, so it is essential to simulate these common issues in order to verify the effectiveness of the test data. If the test data fails to uncover these common errors and omissions, then there are certainly quality issues that need further improvement.

Program mutation means, based on well-defined mutation operations, the program is modified to obtain a mutation program. A well-defined mutation operation can be a simulation of a typical application error. Such as simulating operator usage errors, writing greater-than-equal as less-than-equal, or forcing specific data to appear in order to effectively test a specific code or a specific situation, such as making every expression equal to 0 in order to test a particular situation. With program segment P1, you can replace ">=" in the program with

">" to produce mutation program P2, as shown in Figure 3-20.

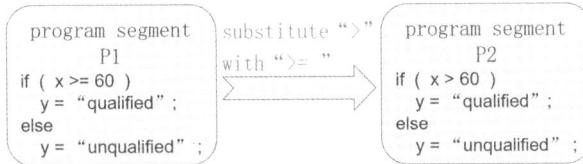

Figure 3-20   Mutation program P2

Program mutation needs to be done under the guidance of the mutation 变异算子
operator. Researchers have proposed a variety of mutation operators, but due to
the different types of programs and their own characteristics, the mutation
operators available at the time of program mutation are also different.

For example, process-oriented programs can be mutated by various
operators mutation, numerical value mutation, method return value mutation and
other operators. For object-oriented programs, while using the above type
mutation operators, new operators can be designed for inheritance,
polymorphism and other features to ensure the completeness of the program
feature coverage. For process-oriented and object-oriented programs, Tables 3-9
and 3-10 list several typical mutation operators respectively. For these mutation
operators, tools such as PITest and MuJava provide good implementations and
support.

Table 3-9   Mutation operators for process-oriented programs

Mutation operator	Description
Operator mutation	(1) For the relational operators "<" "<=" ">" ">=", such as replacing "<" with "<="
	(2) Replacement for the self-increment operator " " or the self-decrement operator "--", e.g., replacing " " with "--"
	(3) Replacement of binary arithmetic operators with numeric operations, e.g., replacing " " with "-"
	(4) Replacement of conditional operators in a program with opposite operators, e.g., replacing "==" with "! ="
Numerical mutation	Taking the opposite of a variable of integer or floating point type in a program, e.g., replacing "i" with "−i"
Method return value mutation	(1) Removing a method in a program whose return value type is void
	(2) Modify the return value of a method in the program, e.g., change "true" to "false"

Table 3-10   Mutation operators for object-oriented programs

Mutation operator	Description
Inheritance mutation	(1) Addition or deletion of overridden variables in subclasses (2) Add, modify, or rename overridden methods in subclasses (3) Remove the keyword super from the subclass, e.g., change "return a*super.b" to "return a*b"
Polymorphic mutations	(1) Instantiating variables as subtypes (2) Change variable declaration, formal parameter type to parent type, e.g. change "Integer i" to "Object i" (3) Replacing the variable used in assignment with other available types
Overloading mutations	(1) Modify the content of the overloaded method, or remove the overloaded method (2) Modify the order or number of method parameters

3.4.2   Mutation test   变异测试

Mutation test, sometimes called mutation analysis, is a technique for evaluating the validity and adequacy of a test data set in order to guide the creation of a more effective test data set. Mutation test was originally developed in the 1970s to locate and reveal deficiencies in software test. If a mutation is introduced, or a known change or even error is inserted into the program, and the test results are not affected, then either the mutation code is not executed, or the program modification or even error is not detected by the test work. Mutated code is not executed, either because there is unreachable code in the source program or because the software is inadequately tested.

There is unreachable code in the source program, which is obviously a problem and needs to be fixed. If the mutated code is executed, but the test results are not affected, it means that the test is not sufficient to find the problem in the program. Mutation test evaluates the error detection capability of a set of test cases by comparing the results of the source program and the mutation program when executing the same test case. When there is a difference, it is considered that the test case can detect the error in the mutation program, and the mutation program is killed; When there is no difference, the test case is considered to have not detected an error in the mutated program, and the mutated program survives.

Execution differences are mainly manifested in the following two scenarios.

(1) When executing the same test case, the source program and the mutation program produce different runtime states;

(2) When executing the same test case, the source program and the mutation program produce different execution results.

Mutation test can be classified into weak mutation test and strong mutation test based on how they meet the requirement for execution difference. In weak mutation test, a mutated program is considered to be "killed" when scenario 1 occurs, whereas in strong mutation test, a mutated program is only deemed "killed" when both scenarios 1 and 2 are satisfied. It is evident that weak mutation test resembles code coverage test and requires less computational power in practice. On the other hand, strong mutation test is more rigorous and can better simulate real error detection scenarios.Before conducting mutation test, it is essential to explicitly specify the type of mutation test and determine the conditions for considering a mutant as "killed". Throughout this book, unless otherwise specified, references to "mutation tests" pertain to strong mutation tests.

Given a program P and a test data set T, generate a set of mutants Mi for P by a mutation operator F (which must be a grammatical change, and the program will still be able to execute after the change), and run a test using T on both P and all Mi. If a Mi produces a different result from P on some test input tj , it is said to be killed; if a Mi produces the same result on all test datasets, then it is said to be a living mutation. Next, living mutations are analyzed to check whether they are equivalent to P. If they are equivalent, they are removed. For mutations Mi that are not equivalent to P, the set of test cases is expanded to improve the error detection capability, and further test is performed. Keep repeating the above process until the test case set can kill all the mutations of the program. The mutation test process can be described as follows. 变异测试过程描述如下

```
Program : P
Test data set: T
Mutation Algorithm: F( )
F( P ) → Mi (i=1, 2, 3...)  // Generate a set of mutations Mi
Test( P, T ) and Test( Mi , T ) // Run a test on P and all Mi using T
```

```
If Test( P, T ) <> Test( Mi , T ) // if test results are different
   Mi is Killed // then the Mi is killed;
Else // otherwise
   Mi is alive // call it a living mutant
Endif
If Mi (alive) <> P Improve ( T ) // If there exists a living
mutation that is not equivalent to P, the set of test cases needs
to be expanded to improve the error detection capability.
```

For program segment P1, we have previously replaced ">=" with ">" to produce the following mutation program P2. In addition, another mutation program P3 can be produced by replacing ">=" with " = ", as shown in Figure 3-21.

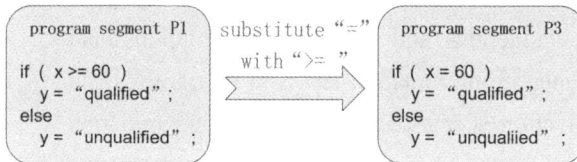

Figure 3-21   Mutation program P3

Supposing a tester A, he designs a test data set T1 for the original program segment P1, including test data x = 70 and x = 50. When this test data set is used for the mutation program P2, no problem can be found, and the two test data both give the correct results, which can remind the tester that it is necessary to increase the number of test cases for T1, such as x = 60, as shown in Figure 3-22.

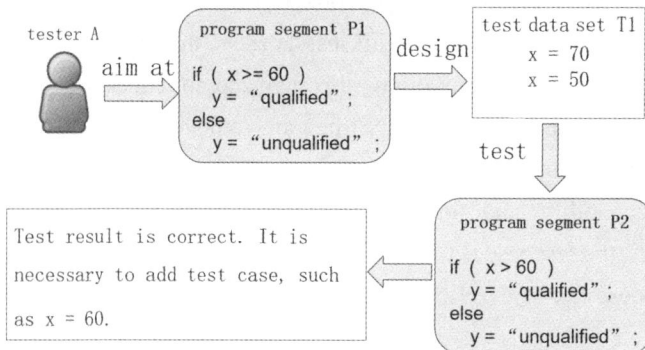

Figure 3-22   Example of test case improvement based on mutation test1

Supposing a tester B, he designs a test data set T2 for the original program segment P1, including test data x = 60 and x = 50. When this test data set is used for the mutation program P3, the problem cannot be found, and the two test data both get the correct result, which can remind the tester that it is also necessary to increase the test cases for T2, as shown in Figure 3-23.

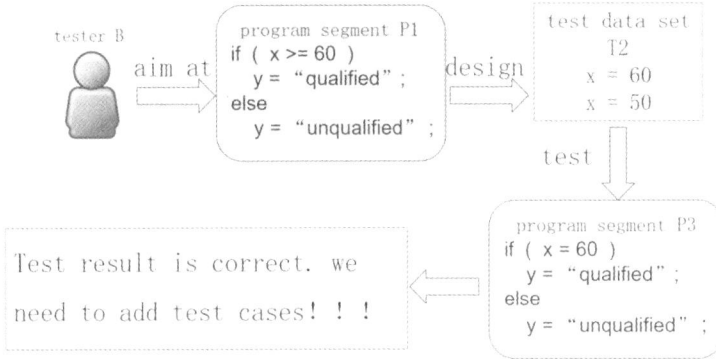

Figure 3-23   Example of test case improvement based on mutation test2

3.4.3   Advantages and disadvantages of mutation test

变异测试的优缺点

From the perspective of software test, mutation test can assist testers in identifying deficiencies in their test work and subsequently improving the coverage, validity, and optimization of the test data set. Additionally, mutation test can be utilized to enhance program source code at a detailed level. Program mutation test is a form of bug-driven test that typically focuses on specific types of program errors. After years of research in test theory and practice, it has become increasingly evident that finding all errors within a program is nearly impossible. A more practical approach involves narrowing the search for errors to facilitate specialized test for specific error types.

In order for mutation test to address a wide range of scenarios, testers must simulate as many potential error scenarios as possible and introduce numerous mutation operators or operations. This results in an extensive number of compiled, executed, and tested mutations programs which consume significant computational resources. As such, it becomes challenging to practically apply this method within current software development processes where version iterations are accelerating rapidly and the cost of test prevents it from becoming a fundamental and widely used software test method. In mutation test, verifying the execution result of a program is also a costly and manual process, which affects the application of mutation test in production practice.

Since equivalent mutation programs are logically undecidable, how to detect and remove equivalent mutation programs from source programs quickly and effectively is also a problem that affects the automation and application of mutation test. Additionally, mutation test relies on having a test data set T, which is designed using other methods, so mutation test generally cannot be used alone. The advantages and disadvantages of mutation test are summarized in Table 3-11.

Table 3-11   Advantages and disadvantages of mutation test

Advantages	Disadvantages
(1) Help to identify deficiencies in the test efforts, increase the coverage and validity of the test dataset, and improve and optimize the test dataset (2) Can be used to improve program source code in terms of details	(1) If mutation test is to be made to target a wide range of situations, a large number of mutations must be introduced, which will result in excessive test costs (2) Mutation test is difficult to automate (3) There is a logical undecidability of the equivalent mutation program (4) Mutation test is generally not used alone, but needs to be combined with other traditional test methodology techniques

Exercise 3

I. Single choice questions.

1. The number of paths contained in a program is directly related to (   ).

    A. the complexity of the program

    B. the number of program statement lines

    C. the number of program modules

    D. the execution time of program instructions

2. The purpose of condition coverage is (   ).

    A. to make each condition in each decision satisfy its possible values at least once

    B. to make each decision in the program get a true and a false value at least once

    C. so that all possible combinations of values for all conditions in each decision occur at least once

    D. so that each executable statement in the program is executed at least once

3. The purpose of software debugging is to (   ).

A. finding hidden errors in software

B. resolving errors found during test

C. minimizing the discovery of errors so that the software can be delivered early

D. proving the correctness of the software

4. There is a program segment as follows.

```
If ((M>0) && (N = = 0))
    FUCTION1.
If ((M = = 10)|| (P > 10))
    FUCTION2.
```

FUCTION1 and FUCTION2 are statement blocks.

Now select the test case:

M = 10 , N = 0 , P = 3

which is satisfied? (   )

A. path coverage   B. condition combination coverage

C. decision coverage   D. statement coverage

5. The test case that satisfies decision coverage for the following program for calculating individual income tax is (   ).

```
if (income<800) taxrate=0.
else if (income<=1500) taxrate=0.05;
else if (income<2000) taxrate=0.08; else taxrate=0.1; else taxrate=0.08; else
taxrate=0.08
else taxrate=0.1.
```

A. income=(799, 1500, 1999, 2000)   B. income=(799, 1501, 2000, 2001)

C. income=(800, 1500, 2000, 2001)   D. income=(800, 1499, 2000, 2001)

6. A program is provided as follows.

```
if (a==b and c==d or e==f) do S1
   else if (p==q or s==t) do S2
        else do S3
```

The minimum number of test cases to achieve "conditional/deterministic coverage" is (   ).

A. 6   B. 8   C. 3   D. 4

Ⅱ. **Fill in the blanks.**

1. Both software test and debugging need to be allowed due to software defects _____.

2. Some defects are not a problem with the program itself, but _____ are amplified or accumulated.

3. _____ is a type of logical coverage criterion that requires that enough test data be selected so that every possible combination of conditions in each judgment expression occurs at least once.

III. True or false questions.

1. The set of all test cases that satisfy the coverage criteria for a combination of conditions also branches the coverage criteria. (   )

2. The purpose of software test is to find errors and correct them.(   )

3. Condition coverage is capable of detecting errors contained in conditions, but sometimes falls short of the coverage requirements for decision coverage.(   )

4. In white-box test, if some kind of coverage reaches 100%, it can be guaranteed that all hidden program defects have been exposed.(   )

5. The condition coverage criterion for white-box test is stronger than the deterministic coverage.(   )

6. Decision coverage includes statement coverage, but it does not guarantee that every error condition can be checked.(   )

IV. Comprehensive question.

1. Design test case sets for the following program segment, which is required to satisfy statement coverage, decision coverage, condition coverage, condition/decision coverage, and condition combination coverage.

```
public int do_work(int A, int B){
    int x=0;
    if((A>4) && (B<9))
    { x = A-B;}
    if( A==5 && B>28 ))
    { x= A B;}
    return x.
    }
```

2. Design test case sets for the following program segment that are required to satisfy separately statement coverage, decision coverage, condition coverage, condition/decision coverage, and condition combination coverage.

```
public void do_work(int x, int y, int z){
    int k=0, j=0; if ( (x>20)&&(z<10) )
```

```
if ( (x>20)&&(z<10) )
{ k=x*y-1;
 j=k*k.
}
if ( (x==22)||(y>20))
{ j=x*y 10; }

j=j%3;
System.out.println("k, j is:" k ", " j);
}
```

3. Design a set of test cases for the following program segment that satisfies the condition combination coverage.

```
public class Triangle {
    protected long lborderA = 0;
    protected long lborderB = 0; protected long lborderC = 0; protected long
lborderC = 0
    protected long lborderC = 0; protected long lborderC = 0; protected long
lborderC = 0
    // Constructor
    public Triangle(long lborderA, long lborderB, long lborderC) {
        this.lborderA = lborderA; this.
        this.lborderB = lborderB; this.
        this.lborderC = lborderC; }


    public boolean isTriangle(Triangle triangle) {
    boolean isTriangle = false; // Check the boundary.
    // check boundary
    if(triangle.lborderA > 0 && triangle.lborderB > 0 && triangle.lborderC > 0)
    // check if subtraction of two border larger than the third
        if ((triangle.lborderA-triangle.lborderB) < triangle.lborderC &&
(triangle.lborderB-triangle.lborderC)   <   triangle.lborderA   &   amp;&
(triangle.lborderA && (triangle.lborderC) > 0 ) // check if subtraction of two
borders larger than the third ) amp;& (triangle.lborderC-triangle.lborderA) <
triangle.lborderB)
        {isTriangle = true; }
    return isTriangle.
    } }
```

4. There is a program module function1 as follows.

```
1   public int Function1(int num, int cycle, boolean flag)
2      {   int ret = 0;
3         while( cycle > 0 )
4           { if( flag == true )
5                { ret = num - 10;
6                  break;
7                }
8           else {
9                if( num%2 == 0 )
10                   { ret = ret * 10;       }
11               else
12                   { ret = ret +1;   }
13               }
14           cycle--;
15           }
16      return ret;
17   }
```

(1) Draw a control flow graph for the program and compute the loop complexity of the control flow graph.

(2) Derive the basic path.

(3) Design the basic path coverage test cases.

5. Please mutate the following code snippet with the mutation rule of replacing " " with "--" and then design the test data to be able to test to find all the mutations.

```
public class zhengchu {
 public String iszhengchu(int n) {
        if(n<100||n>200) {return "error";}
        int flag=0;
        String note=""; if(n%3==0) {return "error";} int flag=0;}
        if(n%3==0) { flag ;
            note=note " 3"; }
        if(n%5==0) { flag ;
            note =" 5"; }
        if(n%7==0) { flag ;
            note =" 7"; }
        return "Divisible by " flag "number, " note;
 }    }
```

6. There is a summation program as follows, where the variable cycle_num is the actual

number of cycles and maxCycleNum is the maximum number of cycles. If the Z-path coverage test and the cyclic boundary condition test of the loop test are used respectively, analyze the value that the formal parameter variable cycle_num should take under these two test methods.

```
int My_Sum(int cycle_num)
{ int i, s, maxCycleNum.
  i = 1; s = 0; maxCycleNum
  s = 0;
  maxCycleNum = 100;
  while ((i <= cycle_num) && (i <= maxCycleNum))
  { s = i.
    i ; }
  return s; }
```

Chapter 4　Dynamic white-box test practice

The test cases generated by dynamic white-box test design must be executed in order to obtain the test results. In the early stage, the software under test needs to be executed manually, then the data of test case is input, and the actual execution result of the software is compared with the expected correct execution result, so as to know whether the test is passed. Nowadays, test cases are typically written into test scripts using test tools, and the test process is carried out by the execution of these scripts.

4.1　Introduction to JUnit unit test

4.1.1　Introduction to JUnit

JUnit 简介

JUnit is an open-source unit test framework for the Java language. It is an instance of the xUnit architecture, and has been widely recognized as the most successful one. JUnit is designed to be compact yet powerful, providing APIs that enable the creation of unit test scripts with clear and reusable results. Additionally, JUnit boasts its own extension ecosystem. The majority of Java development environments have integrated JUnit as a fundamental unit test tool.

4.1.2　Quick start

快速开始

The process of dynamic white-box test of source code can be simply understood as: given parameters, the code under test is called, and then the test results are compared with the expected results. What parameters are given needs to be carefully designed in order to find as many problems and errors as possible with as few tests as possible, so as to improve test efficiency. To call the code under test, you only need to give the object name and method name. The assert class in JUnit provides a series of assertion methods to check that the actual return value of the method under test is consistent with the expected result. AssertEquals (expected, actual) is the most commonly used assert class. The parameter expected is the expected value, and the parameter actual is the actual value after the test is executed. AssertEquals (expected, actual) is used to compare the two values. An example is as follows.

```
assertEquals(0, new Calculate().Subtract(3, 3))
```

This example code gives the parameter (3, 3) and calls Calculate( ).Subtract( ) to test whether the actual execution result is equal to 0. Let's take a quick look at a simple example to get you started with JUnit test. Calculate class is shown as below.

```
public class Calculate
{     public int Add(int a, int b )
      {          return a+b;          }
      public int Subtract(int a, int b )
      {          return a-b;          }
   public int Multiply(int a, int b )
      {          return a*b;          }
   public int Divide(int a, int b )
      {          return a/b;          }
}
```

(1) Select Calculate.java, click the right mouse button, select "New", "Other... " , as shown in Figure 4-1.

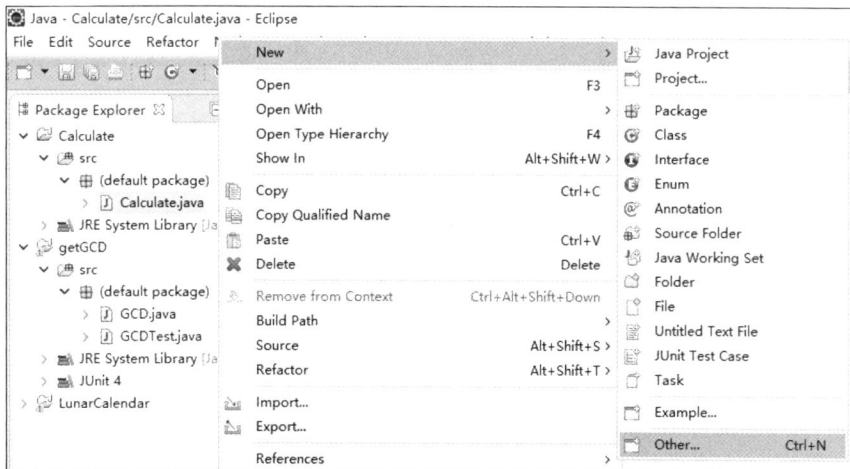

Figure 4-1　Select "New"→"Other..." menu in turn

(2) Select "Java"→"JUnit"→"JUnit Test Case" in the pop-up dialog box, and then click "Next" button, as shown in Figure 4-2.

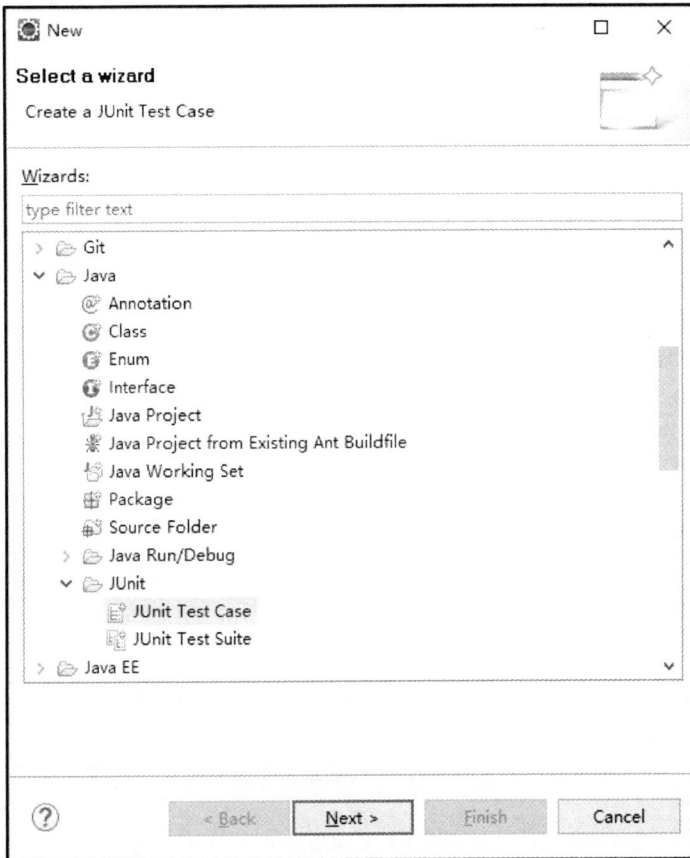

Figure 4-2   Select "JUnit Test Case" in "New" dialog box

(3) Enter and set relevant information in the pop-up dialog box, as shown in Figure 4-3. In this example, the default can be used, and then click "Next" button.

(4) Check "Calculate" class in the dialog box, this will check all the methods of the "Calculate" class by default, there are 4 methods in this example, then click "Finish" button as shown in Figure 4-4.

(5) In this way, we can get the test code structure as shown in Figure 4-5.

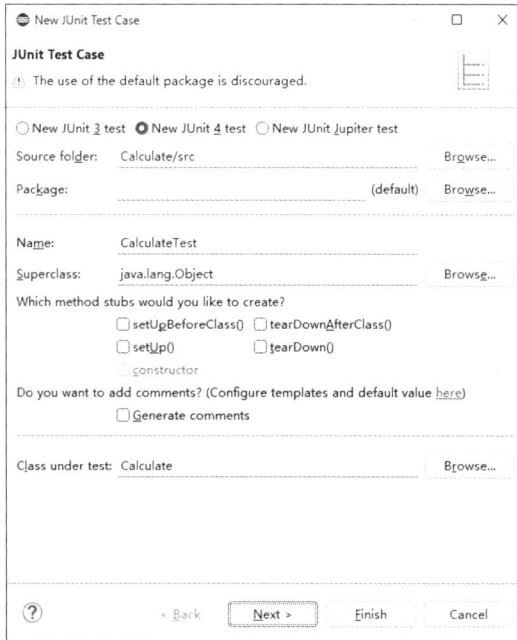

Figure 4-3　Enter and set relevant information

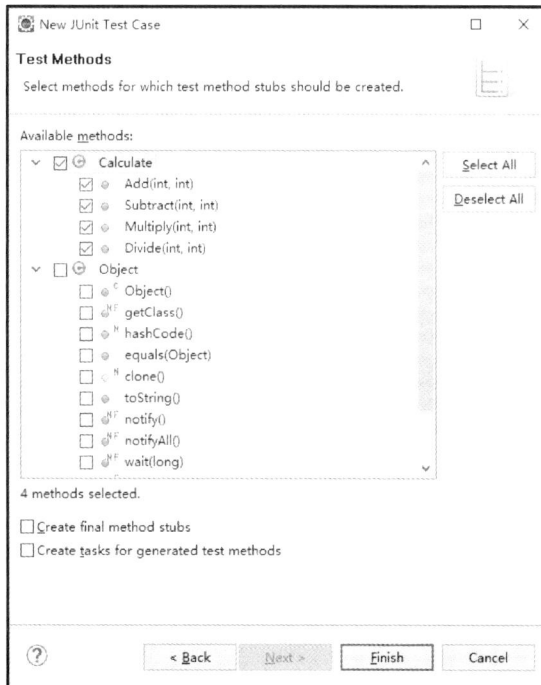

Figure 4-4　Check "Calculate" class

```
 J Calculate.java      J CalculateTest.java ⊠                          ⊟ ⊟
  1⊖ import static org.junit.Assert.*;
  2
  3  import org.junit.Test;
  4
  5  public class CalculateTest {
  6
  7⊖     @Test
  8      public void testAdd() {
  9          fail("Not yet implemented");
 10      }
 11
 12⊖     @Test
 13      public void testSubtract() {
 14          fail("Not yet implemented");
 15      }
 16
 17⊖     @Test
 18      public void testMultiply() {
 19          fail("Not yet implemented");
 20      }
 21
 22⊖     @Test
 23      public void testDivide() {
 24          fail("Not yet implemented");
 25      }
 26
 27  }
```

Figure 4-5   Resulting test code structure

The statement "fail("Not yet implemented"); " indicates that the test script is not complete. The tester needs to write his own test code here; otherwise, an error occurs when the test is executed.

(6) Delete the first "fail("Not yet implemented");" statement, and enter the following code in the corresponding place.

```
assertEquals(6, new Calculate().Add(3, 3));
```

This means that given the argument (3, 3), call Calculate().Add(), and then compare the actual execution to see if the result is equal to 6.

Then click the green execute button and execute the test script CalculateTest.java as shown in Figure 4-6.

(7) You can view the test result after the execution, as shown in Figure 4-7.

This test example runs four tests with 0 Errors and 3 Failures, the Add method passed, and the other three methods failed because there was no corresponding test code.

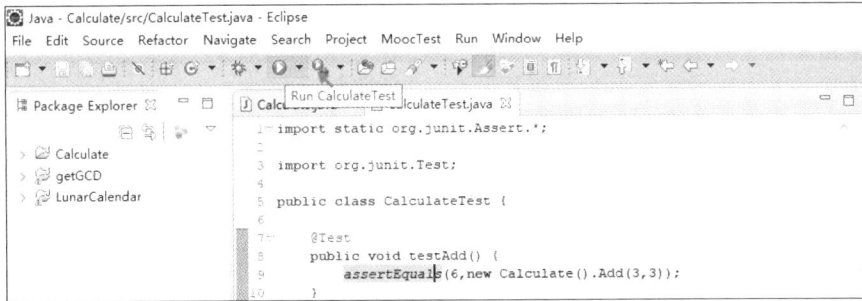

Figure 4-6   Executing the test script CalculateTest.java

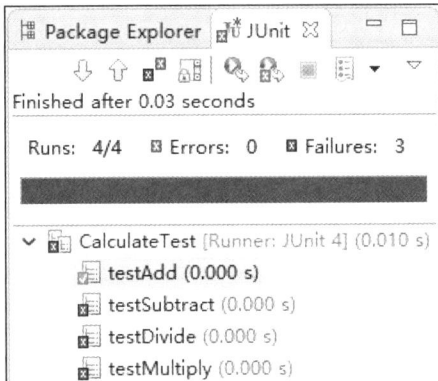

Figure 4-7   Viewing the test result

4.2   Logical coverage example

逻辑覆盖实例

4.2.1   Source code and test requirements

源代码和测试要求

The source code segment is shown as follows. Please design a test script to implement statement coverage and condition coverage test for getGCD method of GCD class.

```
public class GCD {
    public int getGCD(int x, int y){
        if(x<1||x>100)
        { System.out.println("Data out of range!");
          return -1;        }
        if(y<1||y>100)
        { System.out.println("Data out of range!");
        return -1;        }
```

111

```
int max, min, result = 1;
if(x>=y)
{    max = x;
     min = y;              }
else
{    max = y;
     min = x;              }
for(int n=1;n<=min;n++)
{  if(min%n==0&&max%n==0)
    {    if(n>result)
            result = n;              }          }
System.out.println("factor:"+result);
return result;          }          }
```

测试脚本

4.2.2 Test scripts

The test script for statement coverage and condition coverage test of getGCD method is shown as follows.

```
import junit.framework.TestCase;
public class GCDTest extends TestCase {
public void testGetGCD() {
    GCD g = new GCD();
    // statement coverage
    assertEquals(-1, g.getGCD(-1, 15));
    assertEquals(-1, g.getGCD(50, -1));
    assertEquals(5, g.getGCD(50, 15));
    assertEquals(5, g.getGCD(15, 50));

    // condition coverage
    assertEquals(-1, g.getGCD(-1, 15));
    assertEquals(-1, g.getGCD(200, 15));
    assertEquals(-1, g.getGCD(50, -1));
    assertEquals(-1, g.getGCD(50, 200));
    assertEquals(5, g.getGCD(50, 15));
  assertEquals(5, g.getGCD(15, 50));
} }
```

4.3　Object-oriented polymorphic test example

面向对象多态测试实例

4.3.1　Source code and test requirements

The source code segment is shown as follows. The code reflects the characteristics of polymorphism in object-oriented programming. The same method of different classes with inheritance relationship may correspond to different implementation codes. This feature makes the workload of software test multiplied.

```java
public class person {
    String name;
    int age;
    public String introduce() {
        return ("I am a person!");  } }


class student extends person {
public student(String name, int age) {
    super.name=name;
    super.age=age;    }
@Override
public String introduce() {
    return ("I am a student!");  } }


class worker extends person {
public worker(String name, int age) {
    super.name=name;
    super.age=age;    }
@Override
public String introduce() {
    return("I am a worker!");  } }
```

4.3.2　Test scripts

测试脚本

The implementation code of the introduce() method of the three classes person, student and worker are not the same, so we need to test separately. The test script is shown as follows.

```
public class personTest {
    @Test
    public void test() {
    people people0 = new people("Mr. Wang", 25);
    assertEquals("I am a person!", people0.introduce());
  person person1 = new student("Mr. Li", 19);
    assertEquals("I am a student!", person1.introduce());
    person person2 = new worker("Mr. Zhang", 30);
    assertEquals("I am a worker!",person2.introduce());    }  }
```

4.4 Example of national college students software test competition

4.4.1 Source code and test requirements

The national college students software test competition has a previous competition question, the source code is shown as follows. It requires the writing of test scripts to achieve branch coverage.

```
public class Triangle {
protected long lborderA = 0;
protected long lborderB = 0;
protected long lborderC = 0;
// Constructor
public Triangle(long lborderA, long lborderB, long lborderC) {
    this.lborderA = lborderA;
    this.lborderB = lborderB;
    this.lborderC = lborderC;    }
/**
  * check if it is a triangle
  *
  * @return true for triangle and false not
  */
public boolean isTriangle(Triangle triangle) {
    boolean isTriangle = false;
    // check boundary
    if  ((triangle.lborderA  >  0  &&  triangle.lborderA  <=
Long.MAX_VALUE)
```

```
                    && (triangle.lborderB > 0 && triangle.lborderB <=
Long.MAX_VALUE)
                    && (triangle.lborderC > 0 && triangle.lborderC <=
Long.MAX_VALUE)) {

            // check if subtraction of two border larger than the
third
            if (diffOfBorders(triangle.lborderA, triangle.lborderB)
< triangle.lborderC
                    && diffOfBorders(triangle.lborderB, triangle.
lborderC) < triangle.lborderA
                    && diffOfBorders(triangle.lborderC, triangle.
lborderA) < triangle.lborderB) {
                isTriangle = true;          }         }
        return isTriangle;     }
   /**
    * Check the type of triangle
    *
    * Consists of "Illegal", "Regular", "Scalene", "Isosceles"
    */
   public String getType(Triangle triangle) {
        String strType = "Illegal";
        if (isTriangle(triangle)) {
            // Is Regular
            if (triangle.lborderA == triangle.lborderB
                    && triangle.lborderB == triangle.lborderC) {
                strType = "Regular";          }
            // If scalene
            else if ((triangle.lborderA != triangle.lborderB)
                    && (triangle.lborderB != triangle.lborderC)
                    && (triangle.lborderA != triangle.lborderC)) {
                strType = "Scalene";          }
            // if isosceles
            else {   strType = "Isosceles";          }     }

        return strType;    }
   /**
    * calculate the diff between borders
    *
    * */
   public long diffOfBorders(long a, long b) {return (a > b) ?
```

```
(a - b) : (b - a);}
    /**
     * get length of borders
     */
    public long[] getBorders() {
        long[] borders = new long[3];
        borders[0] = this.lborderA;
        borders[1] = this.lborderB;
        borders[2] = this.lborderC;
        return borders;    }  }
```

4.4.2 Test scripts   测试脚本

Test scripts to implement branch coverage of source code are shown as follows.

```
package net.mooctest;
import static org.junit.Assert.*;
import org.junit.Test;

public class TriangleTest {
    @Test
    public void testIsTriangle() {
        assertEquals(false,new Triangle(3,2,1).isTriangle(new
Triangle(3,2,1)));
        assertEquals(true,new  Triangle(3,4,5).isTriangle(new
Triangle(3,4,5)));
        assertEquals(false,new Triangle(-1,4,5).isTriangle
(new Triangle(-1,4,5)));
        assertEquals(false,new Triangle(1,-4,5).isTriangle
(new Triangle(-1,-4,5)));
        assertEquals(false,new Triangle(1,4,-5).isTriangle
(new Triangle(1,4,-5)));
    }

    @Test
    public void testGetType() {
        assertEquals("Scalene",new Triangle(6,8,10).getType
(new Triangle(6,8,10)));
        assertEquals("Regular",new Triangle(2,2,2).getType
(new Triangle(2,2,2)));
```

116

```
        assertEquals("Illegal",new Triangle(3,1,2).getType
(new Triangle(3,1,2)));
        assertEquals("Isosceles",new Triangle(3,2,2).getType
(new Triangle(3,2,2)));
    }

    @Test
    public void testDiffOfBorders() {
        assertEquals(0,new Triangle(2,2,2).diffOfBorders(2, 2));
        assertEquals(1,new Triangle(2,3,2).diffOfBorders(2, 3));
        assertEquals(1,new Triangle(3,2,2).diffOfBorders(3, 2));
    }

    @Test
    public void testGetBorders() {
        Triangle T = new Triangle(2,2,2);
        long[] a = new long[]{2,2,2};
        assertArrayEquals(a,T.getBorders());
    }   }
```

Exercise 4

I. Single choice questions.

1. JUnit's features do not include(   ).

    A. compact size

    B. powerful function

    C. having its own extended ecosystem

    D. most Java development environments that do not integrate JUnit and require additional installation

2. For basic path coverage test of a program, the minimum number of test cases required is(   ).

    A. the total number of paths in the program

    B. the number of decision nodes in the program

    C. number of modules in the program

    D. the loop complexity of the program

3. Which of the following test methods is not a white-box test technique?(   )

    A. Basic path test

B. Boundary value analysis test

C. Programmed piling

D. Logical coverage test

II. Fill in the blanks.

1. The_____ class in JUnit provides a series of assertion methods to check whether the real return value of the method under test is consistent with the expected result.

2. JUnit is an open source Java language_____ framework.

3. The dynamic white-box test process of code can be simply understood as giving _____ , calling the code under test, and then comparing whether the test result is consistent with the expected result.

III. True or false questions.

1. The first parameter of assertEquals( ) is the actual execution result and the second parameter is the expected result.(   )

2. The test cases obtained by the dynamic white-box test design need to be actually executed before the test results can be obtained.(   )

3. Condition coverage is stronger than branch coverage.(   )

IV. Comprehensive question.

1. There is source code snippet.

```
public class boll {
    String boll_color="";
      public String introduce() {
        return ("I am a boll!");    }    }

    class basketboll extends boll {
      public basketboll(String boll_color) {
        super.boll_color=boll_color;        }
      @Override
      public String introduce() {
        return ("I am a " + boll_color + " basketboll" );    }    }

    class footboll extends boll {
      public footboll(String boll_color) {
        super.boll_color=boll_color;        }
      @Override
      public String introduce() {
        return("I am a " + boll_color + " footboll");    }    }
```

(1) Please analyze what this code demonstrate about object-oriented programming.

(2) Please design a test script for the above code.

2. Please mutate the following code snippet, the mutation rule is to replace "++" with "−−", and then design the test script, so that the test can find out all the mutation points.

```java
public class zhengchu {
    public    String iszhengchu(int n) {
        if(n<0||n>500) {
            return "error";
        }
        int flag=0;
        String note="";
        if(n%3==0) {
            flag++;
            note=note+" 3";
        }
        if(n%5==0) {
            flag++;
            note+=" 5";
        }
        if(n%7==0) {
            flag++;
            note+=" 7";
        }
        return "can be "+flag+" number division,"+note;
    }
}
```

Chapter 5　Black-box test design

5.1　Overview of black-box test

黑盒测试概述

Black-box test refers to the approach of treating the software under test as a black box that can't be opened, and the internal logical structure and features of the software are not considered.

According to the specification of the software, the black-box test runs the software, inputs the test data, and checks whether the running results meet the specification.

Black-box test is a kind of test based on specifications from the point of view of users. Black-box test is also known as data-driven test.

5.1.1　Characteristics of black-box test

黑盒测试的特点

Black-box test, which focuses on the execution results and external characteristics of the software without considering its internal structure and implementation details, is commonly used for test the software as a whole, such as system test and acceptance test (as shown in Figure 5-1).

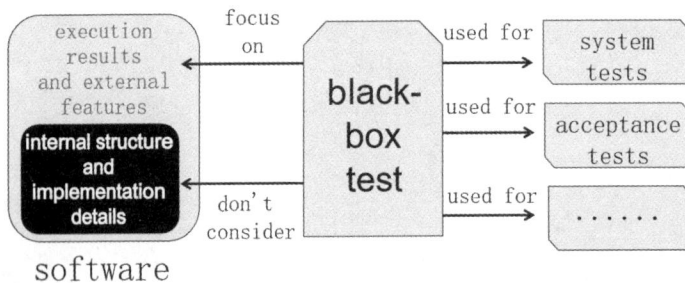

Figure 5-1　Characteristics and uses of black-box test

The design of black-box test cases can be conducted concurrently with software requirements analysis and design, thereby reduces the time required for the entire software project. For instance, during software requirements analysis, preparation for acceptance test and black-box test cases design for acceptance test can be carried out. Similarly, during system specification determining, preparation for system test and black-box test cases design for system test can also be carried out.

The primary foundation of black-box test is software specification. Therefore, prior to initiating black-box test, it is essential to ensure that the software specification undergoes a thorough review and meets the established quality requirements. If there is no specification, exploratory test can be adopted.

Black-box test can be used not only to test the function of software, but also to test the non-functional characteristics of software, such as performance, security, etc. For example, in preparation for the 2022 Winter Olympics in China, Shandong Inspur Ultra HD video industry Co., LTD dedicated 10 months to extensive research and successfully overcame obstacles in 8K Ultra HD live broadcasting technology. This enabled them to offer 8K Ultra HD decoder and video services for live broadcasting of the opening ceremony of the Winter Olympics.

8K is a kind of ultra-high-definition resolution, but the current domestic urban outdoor large screen only supports 4K resolution. In order to ensure the visual experience, the technicians decoded the video into four 4K signals, and then seamlessly spliced and synchronized the four parts of the picture.

Before the opening ceremony, technicians conducted extensive tests on large-screen circuits and control software, striving for that four images look almost perfectly in sync. This requires compatibility test on a variety of large-screen devices and products from mainstream splicer manufacturers in the market. The whole team of more than 40 people repeatedly tested, and the stress test alone was carried out tens of thousands of times.

5.1.2　Main black-box test methods

主要的黑盒测试
方法

The methods for black-box test case design mainly include equivalence class division, boundary value analysis, error guess, cause-effect diagram, decision table driven test, orthogonal experiment design, scenario method, and so on. When facing actual software test tasks, using only one black-box test case design method is insufficient to obtain comprehensive test cases. A practical approach is to utilize a combination of test case design techniques in order to improve both the efficiency and coverage of test. This necessitates a thorough understanding of the principles behind these methods and the accumulation of substantial software test experience in order to effectively elevate the level of software test.

5.1.3　Software defects targeted by black-box test

黑盒测试针对的
软件缺陷

Black-box test can mainly find the following types of errors.

1. Input and output errors

For example, in the user registration interface of an application, there is a text box for entering the user's cell phone number. However, during test it was discovered that the application does not validate the input for a valid cell phone number as required by program specifications. This means that letter input is accepted, which deviates from the intended functionality of receiving a valid cell phone number to send an authentication code later. This failure to perform necessary validation on input data does not align with program specifications and will impact subsequent functions.

2. Initialization, termination error

Initialization error refers to the inability to open the application software normally, as shown in Figure 5-2. A compatibility test of an APP program found that in a particular environment, the APP program will prompt "Security initialization failed. Please start again.", when it is opened after installation.

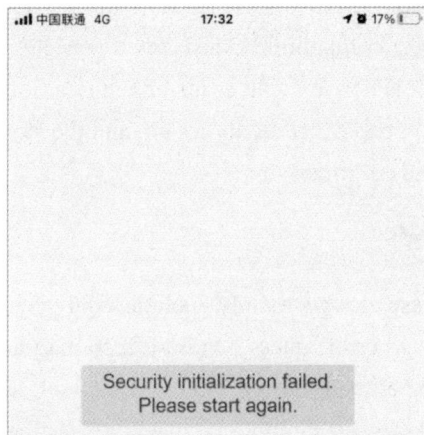

Figure 5-2　APP initialization failure

Termination errors, such as the APP program always being in a running state after test and execution, but no longer responding to user operations without prompts and unable to exit normally.

3. Incorrect or missing functions

The specification of a student schedule query APP states that it can query the schedule of the current teaching week for students based on their student number. However, black-box test reveals that it can only query the weekly schedule for administrative classes and some elective classes cannot be found, indicating a function omission.

4. Interface Error

During black-box test of a grade management APP, the main interface displays "Welcome to the online bookstore," which is an information error in the program's interface.

5. Performance does not meet the requirements

For example, a ticketing APP specifies that it should be able to handle 100,000 mobile customers buying tickets simultaneously. However, black-box test revealed that when simulating 50,000 mobile customers buying tickets at once, the system became paralyzed. This indicates that the system's performance does not meet requirements.

6. Database or other external data access errors

For example, an application requires access to the underlying database during execution, and during black-box test, it encounters a failure in data retrieval. This indicates a database or other external data access error, as illustrated in Figure 5-3. The cause of this error could be attributed to a weak network connection, system congestion, or programming errors.

7. User privacy, security issues, etc.

As an example, black-box test of a student management application revealed that upon logging into the system with a specific student account, it was possible to view and modify the information of other students. Such a system presents user privacy and security issues that may result in unauthorized information disclosure and tampering.

Figure 5-3　Database or other external data access errors

5.2　Equivalence class division test

等价类划分测试

Theoretically, black-box test can theoretically identify all errors in a program by exhaustively test all possible inputs. This includes not only test all legitimate inputs, but also examining those inputs that are not legitimate but likely to occur. However, exhaustive test is impractical, so it is essential to enhance the focus of the test. This involves not only test a variety of potential scenarios to improve test completeness, but also avoiding duplication and reducing redundancy in order to save on test costs. The equivalence class division test represents one such method for black-box test.

5.2.1　Equivalence class division

等价类划分

What is equivalence class division? Let's consider an example. A school requires uniforms for all students. The school uniform factory takes a sample of the uniforms and asks students to try them on. If there are many students in the school, having everyone try on the uniforms can be very time-consuming and laborious. An efficient approach would be to divide the students into different groups based on their body types, as illustrated in Figure 5-4. In this way, only one student from each group needs to try on the uniform. If it fits one student in a group, then it will also fit the other students in that group because they share the same size. This concept is known as equivalence class division.

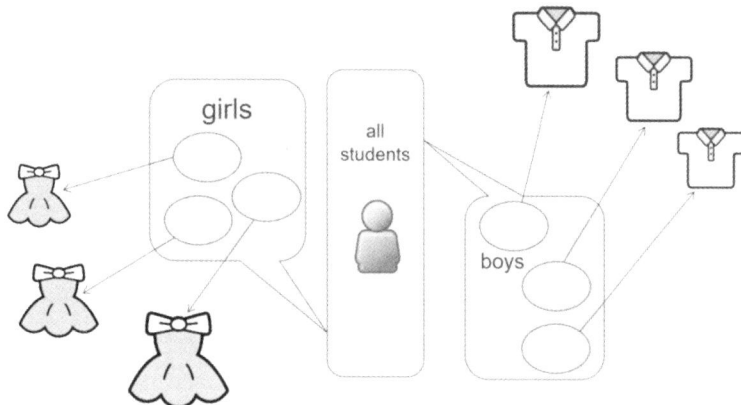

Figure 5-4   Example of equivalence class division

The equivalence class of an element refers to the set of all elements that are equivalent to each other based on a specific equivalence relation. It is a subset of the complete data set, with elements sharing similar characteristics. For instance, integers can be divided into two equivalence classes, odd and even, based on parity.

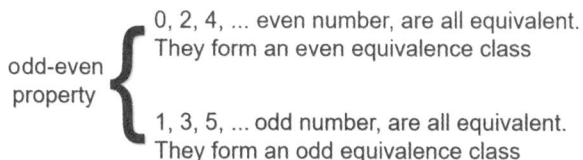

When dividing into equivalence classes, it is important to ensure that there are no duplicate elements between two classes and that when combined, all equivalence classes form the entire data set being divided, as depicted in Figure 5-5.

From the perspective of software test, the equivalence class division method assumes that elements within the same equivalence class share similar characteristics and play an equivalent role in discovering or exposing defects in

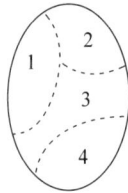

Figure 5-5   Equivalence class division

the program. Therefore, it is reasonable to assume that test one representative data from a certain equivalence class is equivalent to test all data within that class.

In software test, the equivalence class division involves categorizing all possible input data into several equivalence classes and then selecting one or a few representative data from each class to test the program, as illustrated in Figure 5-6.

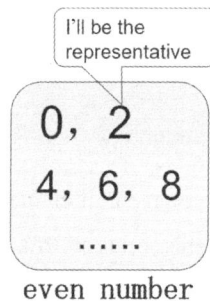

Figure 5-6   For an equivalence class, only a representative of it need be selected for test

By utilizing equivalence class division, we are able to transform potentially infinite inputs into a finite number of equivalence classes. From there, we can select a representative as a test case in order to achieve comprehensive test while minimizing redundancy, reducing costs, and enhancing the effectiveness of the test. Equivalence class division is considered to be the most fundamental and widely used black-box test method.

Equivalence class division test is typically applied to input data based on software specifications. The input data is categorized into different equivalence classes according to various processing methods, and representatives are then chosen from these classes for use as test cases. However, there are instances where equivalence class division test may also be applicable to output data or intermediate process data.

Equivalence classes can be categorized into valid equivalence classes and invalid equivalence classes. A valid equivalence class refers to a collection of input data that is reasonable and meaningful for the program specification. It enables the verification of whether the program meets the predefined features, such as functionality and performance, outlined in the specification. On the other hand, an invalid equivalence class consists of input data that is unreasonable and irrelevant to the program specification. It allows for test whether the program can appropriately handle unusual inputs without undesirable consequences.When designing test cases, it is crucial to consider both types of equivalence classes. This is because software should not only be capable of receiving and processing reasonable data but also resilient enough to withstand unexpected inputs. When faced with unreasonable or irrelevant data inputs, it is essential for the software to handle them properly.

Let's look at an example of the simplest equivalence class division. The symbolic function input x, output y, if $x>0$, then $y=1$; if $x=0$, then $y=0$; if $x<0$, then $y=-1$.

$$\begin{cases} x>0 \to y=1 \\ x=0 \to y=0 \\ x<0 \to y=-1 \end{cases}$$

It is not difficult to classify equivalence classes for x. There are three valid equivalence classes for x, $x>0$, $x=0$ and $x<0$.

And the invalid equivalence classes for x can be categorized as all data that cannot be compared with 0.

In the example of the symbolic function, the valid equivalence class of x is divided according to the range, and for different data types and processing rules, the division of equivalence class is not in the same way. The common ways of division are as follows.

① By range.

② By value.

③ By set.

④ By restriction or restriction rule.

⑤ By processing mode, etc.

For example, to test the individual income tax calculation software, in accordance with the individual income tax classification calculation standards, the input data "annual taxable income" is divided into equivalence class by the range, shown as Table 5-1.

Table 5-1   Equivalence class division of "annual taxable income" according to ranges

Equivalence class number	Annual taxable income	Tax rate /%	Quick deduction
1	Not exceeding 36,000 yuan	3	0
2	Exceeding 36,000 to 144,000 yuan	10	2,520
3	Exceeding 144,000 to 300,000 yuan	20	16,920
4	Part exceeding 300,000 to 420,000 Yuan	25	31,920
5	The portion exceeding 420,000 to 660,000 yuan	30	52,920
6	Over 660,000 to 960,000 yuan	35	85,920
7	Exceeding 960,000 yuan	45	181,920

There is a grade processing program for converting five-level points to a 100-point scale. When testing it, the input data can be divided into equivalence classes according to the processing method, shown as Table 5-2.

Table 5-2   Equivalence classes can be divided according to the processing method

Equivalence class number	Five-level points	Processing
1	Excellent	convert to 90
2	Good	convert to 80
3	Moderate	convert to 70
4	Pass	convert to 60
5	Fail	convert to 40

Up to this point, there is no standardized method for classifying equivalence classes with high quality. Different specifications of the software may require different equivalence classes, and the quality of test cases obtained from these classes can vary. When classifying equivalence classes, the following suggestions can be considered.

(1) If the input condition specifies a range of values, then one valid equivalence class and two invalid equivalence classes can be identified.

For example, if the program input condition is an integer x less than or equal to 100, greater than or equal to 0, the valid equivalence class would be 0

$\leqslant x \leqslant 100$ and the two invalid equivalence classes would be $x < 0$ and $x > 100$.

(2) If the input condition specifies a set of input values, then one valid equivalence class and one invalid equivalence class can be determined.

For example, if a program specifies that valid values for the input data job title come from the set $R = \{$ Teaching Assistant, Lecturer, Associate Professor, Professor, Other, None $\}$, then the valid equivalence class is that the job title belongs to R and the invalid equivalence class is that the job title does not belong to R.

(3) A valid equivalence class and an invalid equivalence class can be determined if the input condition specifies that the input values must satisfy some requirement.

For example, if a program specifies that the input data x must take a numeric symbol as a condition, then the valid equivalence class is x is a numeric symbol and the invalid equivalence class is x contains a non-numeric symbol.

(4) In the case where the input condition is a boolean quantity, a valid equivalence class and an invalid equivalence class can be determined.

For example, if a program specifies that its valid input is a boolean truth value, the valid equivalence class is the Boolean true value, and the invalid equivalence class is the boolean false value.

(5) If the input data is specified as a set of values (assume n) and the program is to process each input values separately, then n valid equivalence classes and one invalid equivalence class can be determined.

For example, the input of a program comes from the set {excellent, good, medium, pass, fail}, and the program will process these 5 values separately, then there are 5 effective equivalence classes, respectively, $x=$ "excellent", $x=$ "good", $x=$ "medium", $x=$ "pass", $x=$ "fail", invalid equivalence class of x is which does not belong to the set {excellent, good, medium, pass, fail}.

(6) If it is stipulated that the input data must conform to certain rules, then it is possible to determine a valid equivalence class (conforming to the rules) and a number of invalid equivalence classes that violate the rules from different perspectives respectively.

For example, a certain message encryption code consists of three parts, the names and contents of which are shown as follows.

Encryption type code : blank or three digits.

Prefix code : a three-digit number that does not begin with "0" or "1".

Suffix code : four digits.

Assuming that the program under test can accept all the information encryption codes conforming to the above regulations and reject all the information encryption codes not conforming to the regulations, using the equivalence class division method, it can be analyzed that all its equivalence classes include 4 valid equivalence classes and 11 invalid equivalence classes, see Table 5-3.

Table 5-3   Example of valid equivalence classes and invalid equivalence classes division

Component	Valid equivalence classes	Invalid equivalence class
Encryption type code	(1) Blank (2) Bit number	(1) With non-numeric characters (2) Less than 3 digits (3) More than 3 digits
Prefix code	(3) 3 digits from 200 to 999	(4) With non-numeric characters (5) Starting bit is "0" (6) Starts with "1" (7) Less than 3 digits (8) More than 3 digits
Suffix code	(4) 4 digits	(9) Non-numeric characters (10) Less than 4 digits (11) More than 4 digits

(7) After the initial division of equivalence, if it is found that there is a difference in the processing of the elements in an equivalence class in a program, the equivalence class should then be further divided into smaller equivalence classes.

5.2.2   Equivalence class division test   等价类划分测试

The process for designing test cases using the method of equivalence class division is as follows.

(1) Divide the equivalence classes, including valid equivalence classes and invalid equivalence classes, establish an equivalence class table, and provide a unique number for each equivalence class.

(2) Design a new test case that covers as many valid equivalence classes as possible that have not already been covered; repeat this step until all valid

equivalence classes have been covered.

(3) Design a new test case so that it covers only one of the invalid equivalence classes; repeat this step until all the invalid equivalence classes have been covered.

As an example, the first step in test the symbolic function is to create a table of equivalence classes, see Table 5-4.

Table 5-4   Creating the equivalence class table

(a) Valid equivalence classes

Input data	Valid equivalence class	Number
x	$x<0$	Y1
x	$x=0$	Y2
x	$x>0$	Y3

(b) Invalid equivalence classes

Input data	Invalid equivalence class	Number
x	Inputs that cannot be compared in size with 0	N1

In the second step, test cases are designed to cover all valid equivalence classes, see Table 5-5.

Table 5-5   Designing test cases to cover all valid equivalence classes

Test case ID	Valid equivalence classes covered	Test data	Expected result
T1	Y1	$x=-4$	$y=-1$
T2	Y2	$x=0$	$y=0$
T3	Y3	$x=8$	$y=1$

In the third step, test cases are designed to cover only one invalid equivalence class at a time, and this step is repeated until all invalid equivalence classes are covered. For symbolic functions, the invalid equivalence classes for the input data x can be grouped together as one class, i.e., all the data that cannot be size-compared with 0, so only one test case needs to be designed to cover it, as shown in Table 5-6.

Table 5-6   Designing a test case to cover the invalid equivalence class

Test Case ID	Valid equivalence class to cover	Test data	Expected result
T4	N1	$x=$"GOOD"	Prompts for an input data error

Why is it that a test case can be designed to cover multiple valid equivalence classes, while only one invalid equivalence class can be covered in general? Supposing there is a grade input software, the input grade consists of two parts, the usual grade cj1 and the final grade cj2, the invalid equivalence classes of cj1 and cj2 are two, less than 0 and greater than 100. The programmer writes the program with two statements to cope with the possible invalid inputs, the code should been as follows.

```
    If ( cj1<0 or cj1>100 ) Return "The usual grade is outside the
range 0-100!"
    If ( cj2<0 or cj2>100 ) Return "The final grade is outside the
range 0-100 !"
```

However, the programmer miswrite cj2>100 as cj2>1000, as follows.

```
    If ( cj1<0 or cj1>100 ) Return "The usual grade is out of the
range 0-100!"
    If ( cj2<0 or cj2>1000) Return "The final grade is out of the
range 0-100!"
```

Now let's design a test case.

cj1 = −10 and cj2 = 800

It covers two invalid equivalence classes, the less-than-0 invalid equivalence class for cj1, and the greater-than-100 invalid equivalence class for cj2.

Run the program, input this test data, after executing the first line of the code to check the validity of cj1, it says "The usual grade is out of the range 0-100!", and exit execution, never continuing with the second line of code at all. So that the error in the second line can not be found and we may also be mistaken that the program successfully passed the test. If we cover only one invalid equivalence class at a time, e.g., cj1 = 70 and cj2 = 800, we can find the error in the second line.

A test case can cover multiple valid equivalence classes, while generally only one invalid equivalence class can be covered, unless covering one invalid equivalence class at a time has been done, specifically to come back to multiple variables to do a combination of invalid equivalence class coverage.

5.2.3   Combination test of equivalence classes   等价类的组合测试

If there are multiple input conditions and a correlation between the conditions, simply covering all the equivalence classes in isolation is not sufficient. It is also necessary to consider the combination of equivalence classes, which can be divided into complete combination and partial combination. When dealing with numerous input conditions and their respective equivalence classes, the total number of complete combinations may become very large. In such cases, partial combination can be utilized.

1. Weak general equivalence class

Design a set of test cases that cover as many valid equivalence classes of the variable under test as possible, ensuring that each valid equivalence class appears at least once. The number of test cases equal the maximum number of valid equivalence classes in each variable under test.

2. Strong general equivalence class

Design a set of test cases that covers all combinations of valid equivalence classes for the variable under test. The number of test cases equal the product of the number of valid equivalence classes for each variable under test.

3. Weak robust equivalence classes

Design a set of test cases, each of which should cover as many valid equivalence classes as possible that have not yet been covered, and at most cover one invalid equivalence class. The number of test cases needed is the maximum number of valid equivalence classes in each variable pluses the total number of invalid equivalence classes in each variable under test.

4. Strong robust equivalence classes

Design a set of test cases that comprehensively cover all combinations of valid and invalid equivalence classes for the variable under examination. The total number of test cases is determined by multiplying the total number of equivalence classes for each variable under examination. The total number of equivalence classes for each variable is calculated as the sum of its valid and invalid equivalence classes.

The meanings of strong and weak, general and robust in the combination of equivalence classes are given below.

$$\begin{cases} \text{weak:at least cover one time} \\ \text{strong:need cover combination} \end{cases}$$

$$\begin{cases} \text{general:only cover valid equivalence class} \\ \text{robust:cover valid and invaild equivalence classes} \end{cases}$$

Let us look at an example below. There is a function $y = f(x1, x2)$. The ranges of value of the input variables are shown as follows.

$x1 \in [a,d]$, $x2 \in [e,g]$

The corresponding equivalence classes are obtained.

$x1$: valid equivalence classes $[a, b)$ $[b, c)$ $[c, d]$, and invalid equivalence classes $(-\infty,a)$, (d, ∞).

$x2$: valid equivalence classes $[e, f)$ $[f, g]$, and invalid equivalence classes $(-\infty,e)$, (g, ∞).

The weak general equivalence class test cases, strong general equivalence class test cases, weak robust equivalence class test cases, and strong robust equivalence class test cases for test function y using equivalence class division are shown in Figures 5-7, respectively.

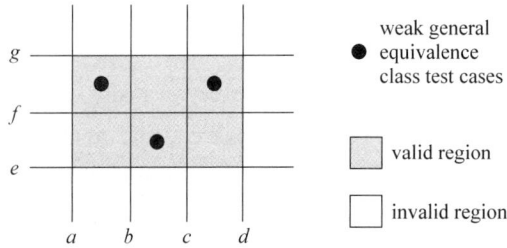

(a) Weak general equivalence class test case

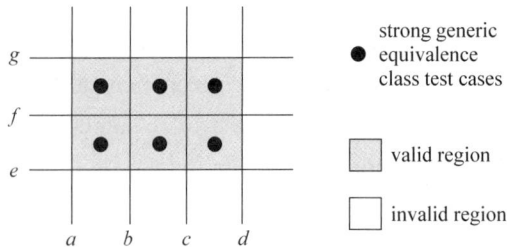

(b) Strong general equivalence class test case

Figure 5-7　Equivalence class combination test case example

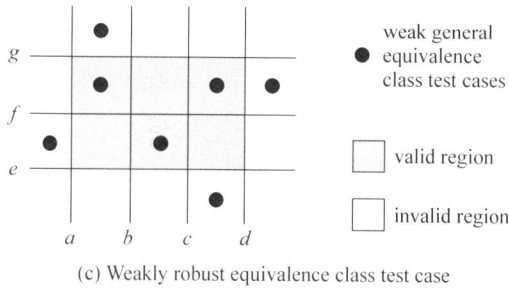

(c) Weakly robust equivalence class test case

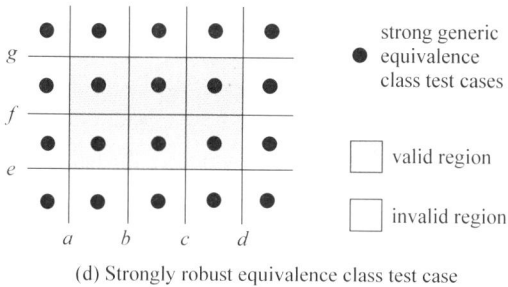

(d) Strongly robust equivalence class test case

Figure 5-7　(continued)

5.3　Boundary value test

边界值测试

People know from long experience in test that a large number of errors tend to occur at the boundaries of input and output data ranges, as shown in Figure 5-8.

Figure 5-8　Errors tend to occur on the boundaries of data ranges

If test cases are designed to cover various boundary cases, it is often possible to uncover more errors. Boundary value analysis is a black-box test method that examines the boundary values of input or output data. It can be used in conjunction with the equivalence class division method, which selects the boundary data of input and output equivalence classes for test based on their division. The key distinction between boundary value analysis and the equivalence class division method lies in the fact that while the latter selects a representative from within an equivalence class at random, the former uses the boundaries of these classes as test conditions.

5.3.1　Boundary value   边界值

To design test cases using the boundary value analysis method, the boundary of the equivalence class should be determined first, and then the values that are exactly equal to, slightly greater than, or slightly less than the boundary should be selected as the test data.

It should be noted that the boundary value can not only be the boundary of the value of the data, but also the number of data, the number of files, the number of records and so on. Typically, there are various types of boundaries that may be targeted by software test, such as: number, character, position, weight, size, speed, orientation, dimension, space, and so on. Accordingly, the boundary value corresponding to the situation may be: the largest/smallest, the first/last, the top/bottom, fastest/slowest, highest/lowest, shortest/longest, empty/full, the leftmost/rightmost, etc.

(1) Edge and corner (leftmost/rightmost/topmost/bottommost) position on the screen.

(2) The first symbol of a string, the last symbol.

(3) The first and last rows of a report.

(4) First and last of an array element.

(5) Loop 1 time, loop maximum.

(6) First record in a data table, last record.

In addition to considering the boundary endpoints themselves, it is important to also consider cases that are slightly greater than or slightly less than these endpoints.

In practical application, the selection of boundary values typically involves the 4-point method and the 6-point method for a given range of values. The 4-point method entails selecting the two endpoints of the value range, as well as one point within each endpoint. On the other hand, the 6-point method involves choosing the two endpoints of the value range, along with one point inside and one point outside each endpoint. This is illustrated in Figure 5-9.

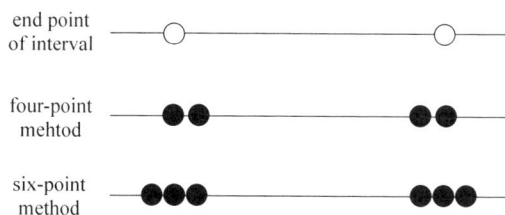

Figure 5-9　4-point and 6-point methods

The 4-point method combined with the normal values in the equivalence class is the 5-point method. The 6-point method combined with the normal value in the equivalence class is the 7-point method. This is shown in Figure 5-10.

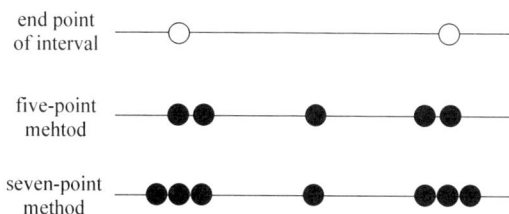

Figure 5-10　5-point and 7-point methods

In most cases, the boundary value can be obtained from the software specification or common sense, however, some boundary values are not directly presented in the software specification, it is easy to be ignored, but also the boundary conditions need to be concerned about the boundary value test, these are called internal boundary value conditions or sub-boundary value conditions.

The main types of internal boundary value conditions are as follows.

(1) Numerical boundary value tests. Computers work on a binary basis, so any numerical operations of the software are limited to a certain range.

(2) Boundary value test for characters. Characters are also important representation elements in computer software, of which ASCII and Unicode are common encoding methods.

(3) Boundary value test for error. Some computational processes have errors, and it is necessary to check whether the error will exceed the acceptable range.

5.3.2　Boundary value test case design

边界值测试用例
设计

The principles of designing test cases with boundary values are as follows.

(1) If the input condition specifies a range of values, the values that are just equal to, slightly greater than, and slightly less than the endpoints of the range should be selected as test input data.

For example, if the specification of the program states, "The formula for calculating postage for mail with a weight in the range of 10 kg to 50 kg is ...". As a test case, we should take 10 and 50, and also 9.99, 10.01, 49.99, and 50.01, etc.

(2) If the input condition specifies the number of values, the maximum number, the minimum number, and the number that is 1 more or 1 less than the maximum number and the minimum number are used as the test data.

For example, an input file should include 1-255 records, then the test case can take 1 and 255, and should also take 0, 2 and 254, 256, and so on.

(3) Use principle (1) for each output condition according to the program specification.

For example, if the specification of a program states that the result of the calculation of the program should be between [0,100], then test cases can be designed such that the expected result of the calculation should be 0, slightly greater than 0, slightly less than 100, and 100.

(4) Use principle (2) for each output condition according to the program specification.

If a program can output up to 5 files at a time, then test cases can be designed such that the expected output is, respectively, 0, 1, 4, and 5 files.

(5) If the specification of a program gives an input domain or an output domain that is an ordered set (e.g., ordered table, sequential file, etc.), then the first and last elements of the set should be selected as test cases.

(6) If an internal data structure is used in the program, the values on the boundaries of this internal data structure should be selected as test cases.

(7) Analyze the program specification to identify other possible boundary conditions.

5.3.3　Combination test of boundary values　　　　　边界值的组合测试

If there are multiple variables, the combination of boundary values of these variables can be categorized into various cases.

1. General boundary values

Consider only the boundary values of a single variable over a valid range of values, including the minimum value, slightly above the minimum value, slightly below the maximum value, and the maximum value. If the number of variables under test is n, there are $4n$ total boundary values. When designing test cases, only the boundary value of one variable is covered at a time, and normal values should be used for the other variables, so a normal value can be selected for each variable, in which case the boundary value and the equivalence class division are combined, and the total number of test cases is $4n+1$.

For example, if program F has two input variables $x1$ ($a{\leqslant}x1{\leqslant}d$) and $x2$ ($e{\leqslant}x2{\leqslant}g$), the general boundary value test cases for F ($x1$, $x2$) takes the following form.

    { <nom ,min>, <nom ,min+ >, <nom ,nom>.

    <nom ,max>, <nom,max−>, <min ,nom>,

    <min+ ,nom>, <max,nom >, <max−,nom> }

Nom denotes the normal value, min denotes the minimum value, max denotes the maximum value, min+ denotes a value slightly greater than the minimum value, and max− denotes a value slightly less than the maximum value. The total number of test cases is $4n+1=4{\times}2+1=9$.

2. General worst boundary values

The combination of boundary values of multiple variables over the valid range is included, using the combination of minimum, slightly above minimum, normal, slightly below maximum and maximum values of each variable as the test cases. If the number of variables under test is n, the total number of test cases is 5^n.

3. Robust boundary values

Considering the boundary values of a single variable on both valid and invalid ranges, in addition to choosing the minimum, slightly above minimum, normal, slightly below maximum and maximum values as boundary values, values that are slightly above the maximum and slightly below the minimum are also chosen. If the number of variables under test is n, the number of test cases is $6n+1$.

4. Robust worst boundary values

Considering the combination of boundary values of multiple variables on both valid and invalid ranges, the boundary values of slightly less than minimum, minimum, slightly more than minimum, normal, slightly less than maximum, maximum and slightly more than large values of each variable are used for the complete combination. If the number of variables under test is n, the number of test cases is 7^n.

The function $y = f(x1,x2)$ with input variables in the ranges of $x1 \in [a,d]$, $x2 \in [e,g]$, has 9 sets of general boundary values as shown in Figure 5-11.

There are 25 sets of general worst boundary values, as shown in Figure 5-12.

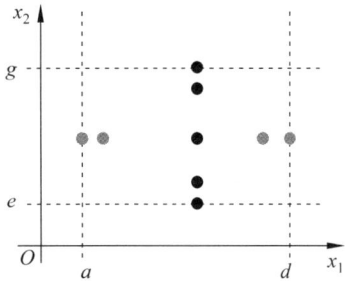

Figure 5-11　General boundary values　　　Figure 5-12　General worst boundary values

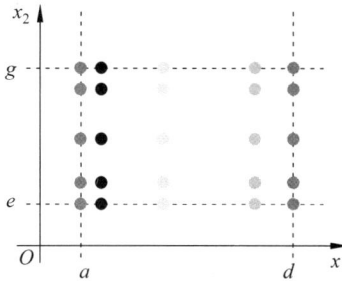

There are 13 sets of robust boundary values, as shown in Figure 5-13.

There are 49 sets of robust worst boundary values, as shown in Figure 5-14.

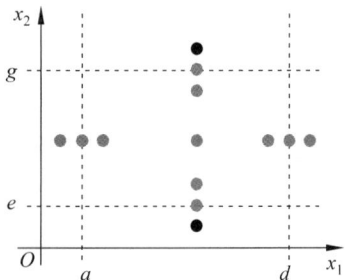

Figure 5-13　Robust boundary values　　　Figure 5-14　Robust worst boundary values

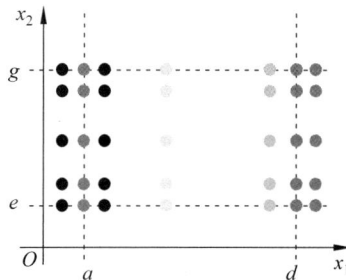

Taking boundary values at the same time on multiple variables looks like a more thorough and better test, but the test cost is really not small. For example, when $n = 3$, the number of test cases to achieve robust boundary value coverage is $6n+1 = 6 \times 3 +1 = 19$, the number of test cases to achieve robust worst boundary value coverage is $7^n = 7^3 = 343$, which is about 18 times as many as the former. When the variables are relatively independent of each other, it is sufficient to use only one variable taking the boundary value and the others taking the normal value, so as to achieve the proper test effect and save a lot of test cost.

5.4   Error guess method

错误推测法

5.4.1   Error guess

Error guess method is that, based on experience, problem analysis or intuition, speculate there may be certain errors in the program, design specific test cases on purpose to test the program, as shown in Figure 5-15.

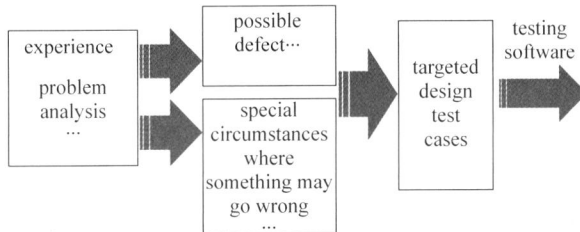

Figure 5-15   Error guess

The basic idea of the error guess method is to list the possible defects in the program, or the special circumstances that may go wrong during the execution of the program, select or design test cases based on them, and then test the program in a targeted manner.

1. Common defects in software

The following are common defects in software.

(1) No restrictions or checks on input data.

(2) There is no constraint on the result set size for a single data query.

(3) The debugging information from the server is displayed on the page when there is an error in the execution of the website page.

2. The situations where errors are likely to occur

The following are the situations where errors are likely to occur during program execution.

(1) Deleting records from an empty data table.

(2) Deleting a record repeatedly.

(3) Adding two identical records.

3. The application of error guess method

Test cases are designed for those common defects in software and the situations where errors are likely to occur, so that problems are likely to be detected. The error guess method for software test necessitates a certain level of experience on the part of testers. This experience enables them to recognize common defects in the software and anticipate potential errors in program execution, thereby facilitating targeted tests.

Testers can leverage their own test practice and accumulated experience to analyze the distribution of software defects, including functional and interface issues. By also learning from others' test experiences, they can compile tables of common software defects and program errors. These tables serve as valuable resources for designing test cases when working with similar software.

Test cases designed using the error guess method have a higher likelihood of uncovering software defects. To effectively utilize this method, testers must not only accumulate experience but also possess a deep understanding of user needs, business processes, and software features related to the system under test. Additionally, they should demonstrate strong problem analysis capabilities and insight—especially when dealing with untested types of software where there is limited prior experience to draw upon. In such cases, testers must rely on their innovative thinking abilities to speculate which parts of the software may not meet user needs or could be flawed.

It's important to note that while the error guess method is effective at identifying defects, it should not be relied upon as the sole test approach. Instead, it should be used as a supplement alongside other test methods due to its inability to guarantee comprehensive test coverage.

5.4.2 Application of error guess for mobile applications

移动应用错误推
测法应用

For mobile applications, the following aspects need to be targeted for test.

(1) Whether the mobile application can run normally on various mobile terminal devices of multiple brands and models with different hardware and software configurations.

(2) When the network bandwidth changes, as well as disconnecting and

reconnecting the network, whether the mobile application will make errors.

(3) Whether the mobile application will make errors when the network environment changes. Changes in the network environment include switching between mobile data and WLAN, etc., and switching between different mobile data types (2G - 5G).

(4) Whether the mobile application will make errors or have problems with the display effect when the screen settings, font settings, etc. of the mobile terminal are changed.

(5) Whether the mobile application can continue to run normally after it is first switched to the background and then switched back to the foreground.

(6) If there is a pop-up window when the mobile application is executed, whether it can continue to run normally.

(7) Can the mobile application continue to run normally if there is an incoming call or SMS when it is executed?

(8) When the mobile application is executing, if the alarm suddenly rings, can it continue to run normally?

(9) Can the mobile application continue to run normally when there are other applications with high memory consumption?

(10) If the installation of the mobile application needs to be verified through the network, you should test whether there is a corresponding prompt for the installation under the circumstance of disconnecting from the network.

(11) If the mobile application is uninstalled in the process of crash, power failure, reboot and other unexpected situations, whether it can be uninstalled correctly after the environment is restored.

(12) When the mobile application is running, using the camera, calculator and other cell phone functions, whether the mobile application can run normally.

5.5   Decision table driven method   判定表驱动法

The equivalence class division method and boundary value analysis method both focus on input conditions, but they do not consider the various combinations of input conditions. In such cases, although the individual input conditions may have been tested, the combination of multiple input conditions may have been ignored. The decision table driven method, on the other hand, focuses on test for various combinations of input conditions.

5.5.1　Decision table

判定表

Decision table is a logical analysis and expression tool, used to analyze and express multiple input conditions under different combinations of values, which different operations will be performed.

For example, there is a "reading guide", it will ask the reader three questions. The reader only need to answer yes or no to each question. The "reading guide" will give reading suggestions based on the reader's answer. Three questions, each question has two answers, then there has a total of 8 (2× 2×2) different combinations of answers. In order to analyze and express these 8 combinations of conditions and the corresponding reading suggestions, a table can be used as shown in Table 5-7.

Table 5-7　Reading guide

Feel tired?	Y	Y	Y	Y	N	N	N	N
Interested in the content?	Y	Y	N	N	Y	Y	N	N
Does the book confuse you?	Y	N	Y	N	Y	N	Y	N
Go back to the beginning of this chapter and reread	✔				✔			
Read on		✔				✔		
Ski p this chapter to the next							✔	✔
Stop and take a break			✔	✔				

Such a table is a decision table. Early in the development of programming, decision tables have been used as an aid in programming. The decision table can express the combination of multiple conditions and complex logical relationships in a clear and specific manner, and can decompose complex problems according to the various possibilities, and then give the operations that should be performed, so as to achieve both simplicity clarity and avoid omissions.

In the program specification, if the specific implementation of different operations depends on a number of different combinations of logical conditions, then you can consider the use of decision tables to analyze and express. A decision table consists of the following four parts, as shown in Figure 5-16.

(1) Condition stub: Lists all the conditions of the problem. The order of the conditions is usually considered irrelevant.

(2) Action stub: lists all possible actions, usually the order in which these actions are listed is not constrained.

(3) Condition items: lists the specific values of each condition.

(4) Action items: list the specific actions that should be taken under the specific values of each condition.

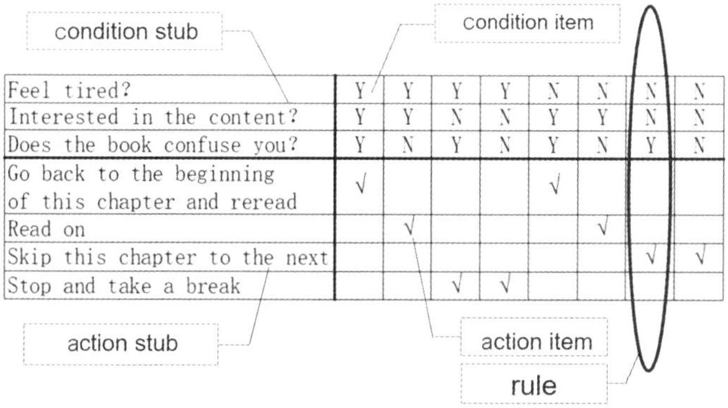

Figure 5-16   Composition of decision table

Each column in the decision table is referred to as a rule. In other words, a specific combination of condition values and their corresponding actions is defined as a rule. A rule consists of specific conditions and action items, which defines the circumstances under which the action takes place. It is evident that there are as many rules as there are different combinations of condition values listed in the decision table. In terms of processing logic, the decision table can break down complex program processing logic into multiple simple processing rules, thereby enabling analysis and understanding of the program, and facilitating easy programming.

Based on the number of condition values, decision tables can be categorized into finite term decision tables and extended term decision tables. A finite term decision table involves each condition having only two values, such as Y/N, T/F, 1/0. On the other hand, an extended term decision table includes conditions with more than two possible values.

5.5.2   Establishment of decision table   判定表的建立

The steps to build a decision table are as follows.

(1) Determine the number of rules.

If there are N conditions and the condition i has M_i values, the total number of rules is as follows.

$$\prod_{i=1}^{N} M_i$$

For example, if a program has 3 input conditions, condition 1 has 2 kinds of values, condition 2 has 4 kinds of values, and condition 3 has 6 kinds of values, the total number of rules is 48 ($2 \times 4 \times 6$).

(2) List all the condition stubs and action stubs.

(3) Fill in the different combinations of values for the conditions.

(4) Fill in the specific actions to get the initial decision table.

(5) Simplify and merge some similar rules with the same action.

Simplification is merging rules. If there are two or more rules that have the same action and their condition terms are similar, then we can consider whether these rules can be combined into one rule, so that the decision table can be simplified.

There is a kind of simplification is more common. For example, a finite decision table has three conditions, including two rules, and the first two conditions take the same value, only one condition takes a different value, but no matter what value this condition takes, the action is the same, which indicates that this condition makes no difference to the results. The two rules can be combined into one rule, the value of the irrelevant conditions can be filled with horizontal lines, as shown in Figure 5-17.

1	2		1
Y	Y		Y
Y	Y	➡	Y
Y	N		—
√	√		√

Figure 5-17　Simplification

What is the use of the decision table for our software test? In fact, each rule in the decision table is a processing logic of the program, for a rule we can design a test case, which is equivalent to test a type of processing logic of the program. When designing a test case for each rule, the condition item constitutes the input of the test case, and the corresponding action item is the expected output result.

5.5.3 Decision table driven test application

判定表驱动测试
应用

A program specification reads, "... gives priority repair treatment to machines with power greater than 50 horse-power and incomplete maintenance records, or machines that have been in operation for more than 10 years. ...". Assuming that "incomplete maintenance records" and "prioritized maintenance" are strictly defined, the following is a 5-step process to create a decision table.

1. Determine the number of rules

There are 3 conditions, and each condition has two values, so there should be 8 ($2\times2\times2$) rules.

2. List all the condition stubs and action stubs

There are three condition stubs: power greater than 50 horse-power, incomplete maintenance records, and has been in operation for more than 10 years.

There are two action stubs: give priority treatment, do something else.

3. Fill in the condition items

There are eight different combinations of condition items, so fill them in the table.

4. Fill in the action items to get the initial decision table

According to the program specifications, fill in the corresponding position in the table with the action that should be performed for each combination of conditions, and then get the initial decision table as shown in Table 5-8.

Table 5-8   Initial decision table

		1	2	3	4	5	6	7	8
Condition	Is it more than 50 horse-power?	Y	Y	Y	Y	N	N	N	N
	Are maintenance records incomplete?	Y	Y	N	N	Y	Y	N	N
	More than 10 years in operation ?	Y	N	Y	N	Y	N	Y	N
Action	Prioritize	✔	✔	✔		✔		✔	
	Other treatment				✔		✔		✔

5. Simplification

After merging similar rules, the final decision table is obtained, as shown in

Table 5-9.

Table 5-9   Decision table after merging similar rules

		1	2	3	4	5
Condition	Is it more than 50 horse-power?	Y	Y	Y	N	N
	Are maintenance records incomplete?	Y	N	N	—	—
	More than 10 years in operation?	—	Y	N	Y	N
Action	Prioritize	✔	✔		✔	
	Other treatment			✔		✔

Next, design five test cases according to the five rules of the final decision table, as shown in Table 5-10.

Table 5-10   Designing test cases based on the decision table

		1	2	3	4	5
Condition	Is it more than 50 horse-power?	Y	Y	Y	N	N
	Are maintenance records incomplete?	Y	N	N	—	—
	More than 10 years in operation?	—	Y	N	Y	N
Action	Prioritize	✔	✔		✔	
	Other treatment			✔		✔
Test case	Input data	Power 80 maintenance record is incomplete	Power 80 maintenance record is complete operation for 12 years	Power 80 maintenance record is complete operation for 5 years	Power 40 operation for 12 years	Power 40 operation for 5 years
	Expected result	Prioritize	Prioritize	Other treatment	Prioritize	Other treatment

5.5.4 The conditions suitable for decision table driven test

The decision table is a concise and easy to be understood tool for analyzing and expressing multi-conditional logic. Of course, it is not always suitable to use decision table driven method to design test cases, and the conditions suitable for using it are shown as follows.

(1) The specification is given in the form of a decision table or can be easily converted into one.

(2) The order in which conditions are listed does not and does not affect which operations are performed.

(3) The order in which the rules are listed does not and does not affect which operations are performed.

(4) Whenever the conditions of a rule have been satisfied and the operation to be performed has been determined, it is not necessary to test other rules.

(5) If more than one operation is to be performed to satisfy a rule, the order in which these operations are performed is irrelevant.

5.6 Cause-effect diagram method 因果图法

A program can be defined as a process that transforms input into output based on specific rules. In this context, the input conditions serve as the cause, while the resulting output or change in program state represents the effect (refer to Figure 5-18). It is important to understand what type of input will yield a particular kind of output, and how different causes lead to distinct results.

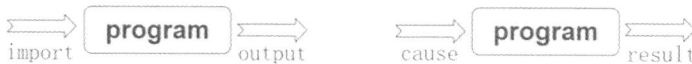

Figure 5-18   Input is the cause, output is the effect

A cause-effect diagram is a valuable tool for illustrating the relationship between various causes and their corresponding effects. This graphical representation allows for a clear depiction of the causal connections among input conditions, constraints, and output outcomes. Cause-effect diagram are commonly utilized alongside decision tables. The cause-effect graph method involves identifying causal relationships from natural language program specifications, representing these relationships using cause-effect diagrams, creating decision tables based on these diagrams, and subsequently generating test cases from the decision table (as depicted in Figure 5-19).

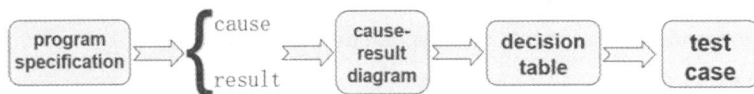

Figure 5-19　Cause-and-effect diagram method

If the decision table can be easily derived from the program specification, there is no need to draw a cause-effect diagram. However, in more complex problems, the cause-effect diagram method is often very effective. For instance, when dealing with a large number of input conditions, using a decision table directly may result in an excessive amount of combinations of conditions, leading to too many columns and increased complexity in the decision table. In reality, there may be constraints between these conditions, rendering many combinations invalid and redundant in the decision table. In such cases, drawing a cause-effect diagram first allows for conscious exclusion of these invalid combinations when subsequently creating the decision table based on the cause-effect diagram. This approach significantly reduces the number of columns in the decision table.

5.6.1　Introduction to cause-effect diagram　　　　因果图介绍

In a cause-effect diagram, C_i is usually used to denote a cause, which is placed in the left part of the diagram; E_i denotes a effect, which is placed in the right part of the diagram. Both C_i and E_i can take the value 0 or 1, 0 means a state does not appear, 1 means a state appears. The cause and effect are connected by a straight line.

1. Relationships

The cause-effect diagram uses 4 types of symbols to represent the 4 types of cause and effect relationships in the program specification, as shown in Figure 5-20.

(1) Equal: If the cause occurs, the effect occurs; if the cause does not occur, the effect does not occur.

(2) Not (~): if the cause appears, the effect does not appear. If the cause

does not appear, the effect appears.

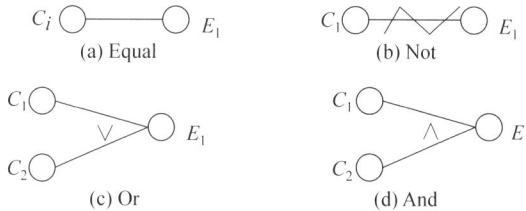

Figure 5-20   4 types of causal relationships

(3) Or ($\vee$): if 1 of several causes appears, result appears; if none of several causes appears, result does not appear.

(4) And ($\wedge$): if several causes all appear, result appears. If 1 of the causes does not appear, the result does not appear.

2. Constraints

Input conditions may have relationships with each other, which is called a constraint. For example, some input conditions cannot occur at the same time. There are also often constraints between output states. In a cause-effect diagram, constraints are labeled with specific symbols.

There are 4 types of constraints for input conditions.

(1) E constraint (mutually exclusive): indicate that there are no simultaneous 1. Such as there is only one 1 at most in a, b, and c.

(2) I constraint (inclusive): indicates at least one 1, i.e., not simultaneously 0 in a, b, c.

(3) O constraint (unique): indicates that there is one and only one 1 in a, b, c.

(4) R constraint (required): indicates that if $a = 1$, then b must be 1. That is, it is not possible for $a=1$ and $b=0$.

The only constraints on the output is M constraint (mask): if the result a is 1, then the result b is forced to be 0.

The 5-class constraint is shown in Figure 5-21.

For program of relatively large size, it is difficult to use a cause-effect diagram as a whole because the number of combinations of input conditions is too large. It can be divided into several parts, and then more detailed cause-effect diagrams can be drawn for each part separately.

151

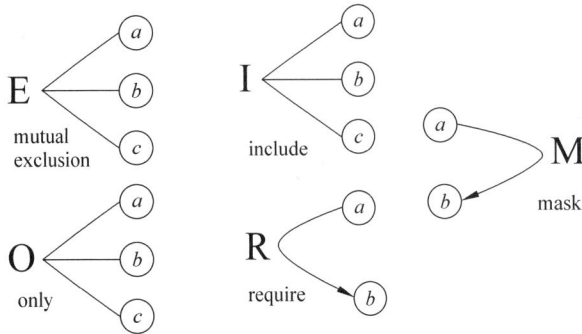

Figure 5-21   5 types of constraints

5.6.2   Steps for designing test cases using cause-effect diagram

采用因果图法设计
测试用例的步骤

The steps for designing test cases using cause-effect diagram method are as follows.

(1) Analyze the description of the software specification to determine what are the causes, i.e., the causes are the input conditions or their equivalence class, and what are the results, i.e., the results are the operations and the outputs, and assign an identifier to each cause and result.

It should be noted that the cause and effect need to be atomized. For example, "the basic salary of the male staff with the job title of engineer is increased by 100, and the bonus is increased by 50", in this software specification description, the cause is two, title = engineer, sex = male; the result is also two: basic salary = basic salary +100, bonus = bonus + 50.

(2) Analyze the semantics in the description of the software specification to find out the relationships between cause and effect, and draw a cause-effect diagram based on these relationships.

(3) Label the constraints. Due to some limitations, some combinations between cause and cause, and some situations between cause and effect are not possible, In order to indicate these special cases, standard symbols should be used to mark the constraints on the cause-effect diagram.

(4) Convert the cause-effect diagram into a decision table.

5.6.3   Test application of cause-effect diagram method

因果图法测试应用

There is a software for a beverage vending machine that handles a unit price of ¥0.5 coin with the following specification. If you put in ¥0.5 coin or ¥1 coins and press the button of 〖orange juice〗 or 〖beer〗, the corresponding drink will be sent out. If the vending machine has no change for give, a red light showing 〖no change〗 will be on, then after putting in ¥1 coin and pressing the button, the drink will not be sent out and the ¥1 coin will be send back. If there is change for give, the red light showing 〖no change〗 will be off, and ¥0.5 coin will be refunded at the same time as the drink is sent out.

The process of designing test cases for this software using the cause-effect diagram method is as follows.

(1) Analyze the specification of this vending machine software and list the causes and effects.

Cause:

① The vending machine has change

② Put in a ¥1 coin

③ Put in a ¥0.5 coin

④ Push the orange juice button

⑤ Push the beer button

Result:

㉑ The light of 〖no change〗 flash

㉒ Return the ¥1 coin

㉓ Return the ¥0.5 coin

㉔ Give out a jar of orange juice

㉕ Give out a can of beer

(2) Draw a cause-effect diagram

The cause is on the left and the effect is on the right, connecting the cause to the effect according to the software specification. There are also intermediate nodes that can be introduced in the cause-effect diagram to indicate intermediate states of processing. The intermediate nodes for this example are as follows.

⑪ Put in a ¥1 coin and press the drink button

⑫ The button has been pressed (〖orange juice〗 or 〖beer〗

⑬ Should be returned ¥0.5 coin and it has change

⑭ Account paid

(3) Constraints are added to the cause-effect diagram, as shown in Figure 5-22.

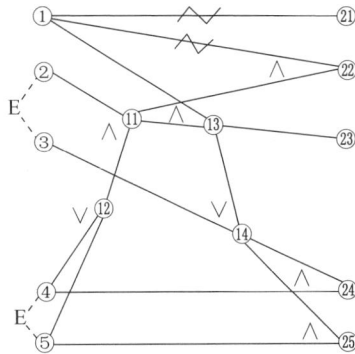

Figure 5-22　Complete cause-effect diagram

(4) Convert the cause-effect diagram into a decision table, as shown in Table 5-11.

Table 5-11　Decision table based on the cause-effect diagram

		1	2	3	4	5	6	7	8	9	10	11	12	13	14	15	16	17	18	19	20	21	22	23	24	25	26	27	28	29	31	31	32
condition	①	1	1	1	1	1	1	1	1	1	1	1	1	1	1	1	1	0	0	0	0	0	0	0	0	0	0	0	0	0	0	0	0
	②	1	1	1	1	1	1	1	0	0	0	0	0	0	0	0	1	1	1	1	1	1	1	1	0	0	0	0	0	0	0	0	0
	③	1	1	1	1	0	0	0	0	1	1	1	1	0	0	0	0	1	1	1	0	0	0	0	1	1	1	1	0	0	0	0	0
	④	1	1	0	0	1	1	0	0	1	1	0	0	1	1	0	0	1	1	0	1	1	0	0	1	1	0	0	1	1	0	0	0
	⑤	1	0	1	0	1	0	1	0	1	0	1	0	1	0	1	0	1	0	1	0	1	0	1	0	1	0	1	0	1	0	1	0
intermediate result	⑪						1	1	0		0	0	0		0	0	0					1	1	0		0	0	0		0	0	0	
	⑫						1	1	0		1	1	0		1	1	0					1	1	0		1	1	0		1	1	0	
	⑬						1	1	0		0	0	0		0	0	0					0	0	0		0	0	0		0	0	0	
	⑭						1	1	0		1	1	1		0	0	0					0	0	0		1	1	1		0	0	0	
result	㉑						0	0	0		0	0	0		0	0	0					1	1	1		1	1	1		1	1	1	
	㉒						0	0	0		0	0	0		0	0	0					1	1	0		0	0	0		0	0	0	
	㉓						1	1	0		0	0	0		0	0	0					0	0	0		0	0	0		0	0	0	
	㉔						1	0	0		1	0	0		0	0	0					0	0	0		1	0	0		0	0	0	
	㉕						0	1	0		0	1	0		0	0	0					0	0	0		0	1	0		0	0	0	

In the decision table above, the shaded portion indicates a situation that is not possible due to violation of constraints and can be deleted.

(5) Design test cases according to the decision table.

5.7　Scenario method

场景法

The black-box test methods introduced in the previous sections are primarily designed for a single function point, and do not encompass the continuous execution of multiple operation steps or the combination of multiple function points. Furthermore, these methods cannot cover the dynamic execution process involving user operations. For complex software systems, it is essential to not only test individual function points but also to understand the entire system from a global business process perspective. This ensures that the program can execute correctly in scenarios involving multiple function points and complex constraints.

The scenario method is a test case design approach that utilizes scenarios to encompass the functional points and business processes of the system, thereby enhancing test effectiveness. This concept was proposed by rational and is detailed in RUP2000's Chinese version with explanations and application examples. By incorporating this approach from software design into software test, it becomes possible to depict situations when events are triggered, facilitating test case design and making them easier to comprehend and execute.

5.7.1　Event flow　　　　　　　　　　　　　　　　　　事件流

Nowadays, software is typically activated by events to regulate the flow. For instance, when applying for a project, one must first submit approval documents, followed by approval from the department manager and ultimately from the general manager. If the department manager does not approve the audit, it will be rejected directly (see Figure 5-23).

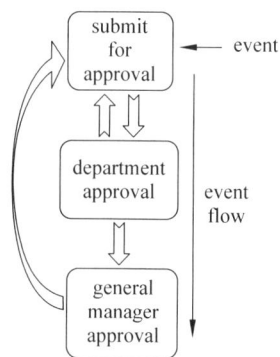

Figure 5-23　Event to trigger the control process

Multiple events triggered in turn form the event flow. In scenario method, the event flow is divided into basic flow and alternative flow. The basic flow refers to the execution path through which each step of the program works "normally". It is the simplest path for program execution, and the program has only one basic flow. Alternative flows refer to paths that a program may or may not pass through. Alternative flows are optional or alternative to the basic flow, and generally correspond to abnormal event flows, as shown in Figure 5-24. A black straight line represents the basic flow and an arc of different colors represents the alternative flow. An alternative flow may start from the base flow, execute under certain conditions, and then rejoin the base flow (for example, alternatives 1 and 3). It may also originate from another alternative flow (such as alternative flow 2), or terminate the use case without rejoining a particular flow (such as alternative flows 2 and 4).

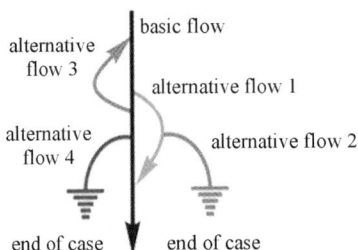

Figure 5-24　Basic and alternative streams

5.7.2　Scenario

场景

Starting from the basic flow, a scenario can be identified by describing the paths passed through. A scenario is an instance of an event flow that corresponds to a sequence of actions performed by the user to execute the software, as shown in Figure 5-25. The scenario method requires traversing the basic flow and all alternate flows.

Scenarios consist of four main types: normal use case scenarios, alternate use case scenarios, abnormal use case scenarios, and speculative special scenarios. The scenario method requires that scenarios be designed to cover all the event flows according to the event flow information contained in the software requirements specification, and that the corresponding test cases be designed so that each scenario occurs at least once, as shown in Figure 5-26.

The steps for designing test cases using the scenario method are as follows.

(1) According to the description, describe the basic flow and alternative flows of the program.

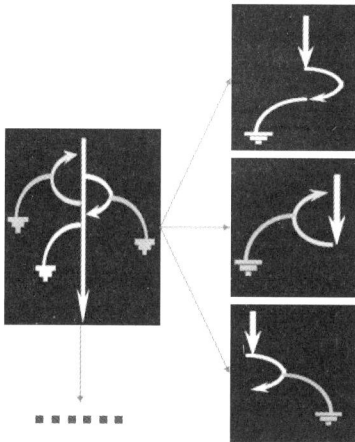

Figure 5-25   Scenario   Figure 5-26   Scenario method

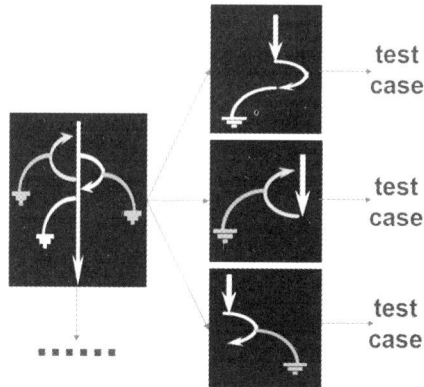

(2) Generate different scenarios based on the basic flow and alternative flows.

(3) Generate test cases for each scenario.

(4) Review all the generated test cases and remove the redundant test cases. After the test cases are identified, determine the test data values for each test case.

In response to Figure 5-24, which shows the event flow diagram of a program, using the scenario method, eight scenarios can be designed to cover the basic flow and each alternative flow.

Scenario 1: Basic flow.

Scenario 2: Basic flow, alternative flow 1.

Scenario 3: Basic flow, alternative flow 1, alternative flow 2.

Scenario 4: Basic flow, alternative flow 3.

Scenario 5: Basic Flow, alternative flow 3, alternative flow 1.

Scenario 6: Basic Flow, alternative flow 3, alternative flow 1, alternative flow 2.

Scenario 7: Basic flow, alternative flow 4.

Scenario 8: Basic flow, alternative flow 3, alternative flow 4.

Note: Scenarios 5, 6 and 8 only consider the case where alternative flow 3 is executed once as a cycle.

In addition to the above eight scenarios, more scenarios can be constructed.

157

The construction of scenario is actually equivalent to the construction of business execution path. The more alternative flows, the more execution paths and the more scenarios. The same alternative flows executed in a different order may form different business processes and execution results. The problem with this is when the number of alternative flows is large, it will lead to scenarios explosion.

The basic principles of how to select typical scenarios for test to meet the requirements of completeness and non-redundancy of test are as follows.

(1) The minimum number of scenarios is equal to the total number of basic flow and alternative flows.

(2) There is one and only one scenario that contains only the basic flow.

(3) For an alternative flow, at least one scenario should cover it, and in that scenario, it should avoid covering other alternative flows as much as possible.

5.7.3　Application of scenario method　　　　　　场景法应用

There is a shopping APP. After the user successfully logs into the system, he/she first chooses the products, then pays for the purchase online, and generates an order after the payment is successful, at last completes the shopping. For such a system, the process of designing test cases using scenario method is as follows.

(1) According to the description, draw the basic flow and each alternative flow of the program , as shown in Figure 5-27.

(2) Generate different scenarios cover the basic flow and all alternative flows.

Scenario 1: Basic flow.

Scenario 2: Basic flow, alternative flow 1.

Scenario 3: Basic flow, alternative flow 2.

Scenario 4: Basic flow, alternative flow 3.

Scenario 5: Basic flow, alternative flow 4.

(3) Generate test cases for each scenario.

Supposing there is a valid account with the username abc, password 123, and account balance 200. The test cases designed for each scenario are shown in Table 5-12.

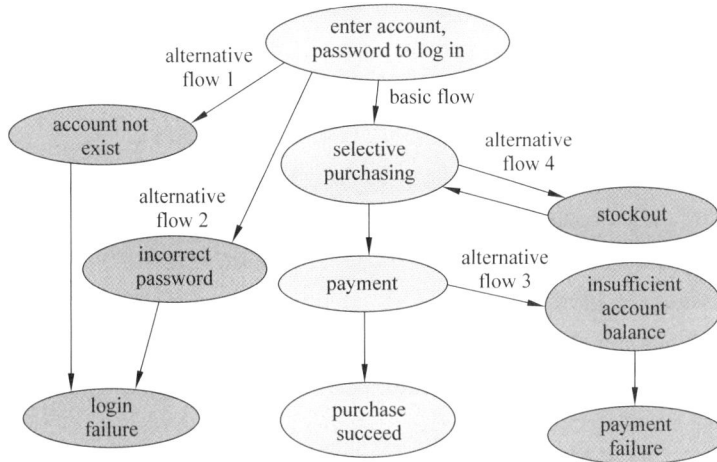

Figure 5-27 Basic flow and alternative flows of the program

Table 5-12 Test cases designed for each scenario

Use case ID	Scenario/ condition	Account number	Pass-word	Operation	Expected result
1	Successful shopping	abc	123	Log in to the system and shop for a product in stock with a price of 50	Payment is successful generate order
2	Account does not exist	aaa (assuming this account does not exist)	123	Login to the system	Login failed
3	Wrong password	abc	345	Login system	Login failure
4	Out of stock	abc	123	Log in to the system and select an out-of-stock product	The product is out of stock
5	Balance is insufficient	abc	123	Log in to the system and purchase a stocked product with a price of 500	The purchase fails because the balance is insufficient

5.8 Orthogonal experiment method

正交实验法

5.8.1 Background of orthogonal experimentation

正交实验的背景

When using cause-effect diagrams to design test cases, it can be challenging to derive the relationship between input conditions and output results from the software requirements specification. The complexity of this cause-effect relationship often leads to a large number of test cases, which can impose a heavy burden on software test. In order to reduce test costs and improve efficiency, the orthogonal experiment design method can be utilized.

For example, when testing the compatibility of an APP, including installation, uninstallation, and various functions (a total of 100 test points), there are several aspects to consider. These include hardware device compatibility, operating system compatibility, resolution and display settings compatibility, network operator compatibility, as well as other software compatibility.

1. Cell phone hardware

Based on current market share data, select the top 15 cell phone brands and models in terms of market share, and then randomly select 5 cell phone brands and models from the remaining. A total of 20 cell phone are selected as the hardware test environment.

2. Operating system versions

There are various versions of android operating system, after research, there are 4 major versions of android operating system.

3. Resolution, display settings compatibility

Different mobile devices have different resolutions, and the same mobile device may also have multiple display settings, assuming that each cell phone brand model may have 4 different resolutions or display settings.

4. Network type

Network types include 2G, 3G, 4G, 5G, WiFi, hotspot, etc., setting a total of 6 types.

5. Compatibility with other software

From the major application market, get the ranking of APP, according to the classified download quantity, selected 6 APPs to do software compatibility test with it.

If we want to test all the test items in the case of complete combination of all the above execution environment elements, the total number of test tasks is as follows.

$$20 \times 4 \times 4 \times 6 \times 6 \times 100 = 1,152,000$$

To test all these situations, the workload is too large. In order to solve this problem, because there are too many possible combinations of conditions, it is difficult to conduct a comprehensive test, orthogonal experiment method can be used. Orthogonal experiment method, is also known as orthogonal design experiment method. The background of its application is as follows: the change of the value of multiple factors will affect the outcome of an event, and it is necessary to verify this effect through experiments; The number of influencing factors is relatively large, and each factor has a variety of values, and the experiment amount is very large. You can't experiment with every possible set of data. Orthogonal experiment method is a kind of experiment design method to select appropriate and representative data from a large number of experimental data for test.

5.8.2 Orthogonal experiment design method

正交实验设计方法

Orthogonal experiment design method is a scientific experiment design method that selects appropriate and representative data from a large amount of (experimental) data and arranges the experiments in a reasonable way. It is based on orthogonality, and selects a part of the representative data from all the data for the experiment. These representative data have the characteristics of "evenly dispersed, neat comparable". It is a kind of high-efficiency, rapid and economical experiment design method.

Experiment workers in the long-term work summed up a set of measure, creating the so-called orthogonal table. According to the orthogonal table to arrange the experiment, not only can make the experiment distribution is very uniform, but also can reduce the number of experiments, and the calculation and analysis is simple, can clearly clarify the relationship between the experimental conditions and results. That the use of orthogonal tables to arrange experiments and analyze the results of experiments, is called orthogonal experiment design method. Simply put, it is the former to summarize the orthogonal table, the latter directly applied to solve specific problems.

In orthogonal experiment method, the conditions that may affect the experimental results are called factors. The number of possible conditional values is called the level (or state) of the factor. An orthogonal table is a set of tables designed with rules, using L as the code name for the orthogonal table, n as the number of experiments to be performed, c as the number of columns, that is, the number of factors affecting the outcome, and t as the number of levels, that is, the number of possible values of the factor. The construction of orthogonal tables requires knowledge of combinatorics and probability theory.

Orthogonal tables of type L_n (t^c) are now widely used, L_4 (2^3) and L_8 ($4^1 \times 2^4$) are shown in Figure 5-28.

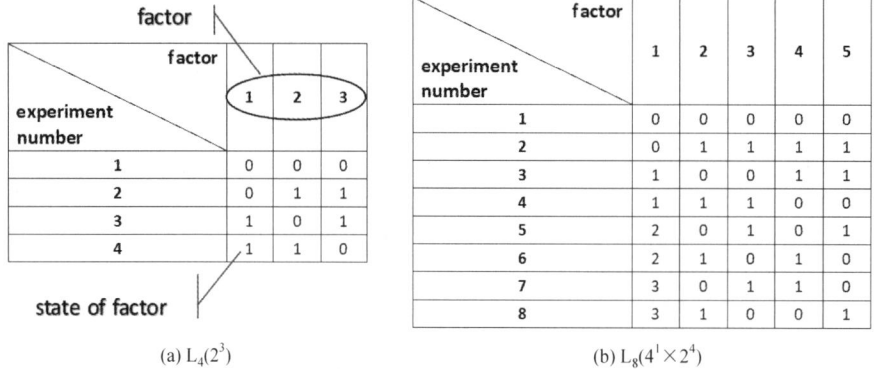

(a) $L_4(2^3)$

experiment number \ factor	1	2	3
1	0	0	0
2	0	1	1
3	1	0	1
4	1	1	0

state of factor

(b) $L_8(4^1 \times 2^4)$

experiment number \ factor	1	2	3	4	5
1	0	0	0	0	0
2	0	1	1	1	1
3	1	0	0	1	1
4	1	1	1	0	0
5	2	0	1	0	1
6	2	1	0	1	0
7	3	0	1	1	0
8	3	1	0	0	1

Figure 5-28　L_n (t^c) type orthogonal table

L_8 ($4^1 \times 2^4$) indicates that if there are one 4-level factor and four 2-level factors, the number of experiments required is 8. Specifically speaking, if there are five input conditions, with four possible values for condition 1 and two possible values for each of condition 2, 3, 4, and 5, the number of tests required is 8.

Instead of using an orthogonal table, if all possible situations are tested, a total of 64 ($4 \times 2 \times 2 \times 2 \times 2 \times 2$) tests are required. The usefulness of orthogonal tables is obvious in reducing the number of experiments.

Orthogonal tables can be categorized into uniform level number orthogonal tables and mixed level number orthogonal tables. The former means that the number of levels of each factor in the table is the same. The latter means that the number of levels of each factor in the table is not the same.

5.8.3 Application steps of orthogonal experiment method

正交实验法应用
步骤

When utilizing the orthogonal experiment method, the conditions of the tested object are considered as the factors in the orthogonal table, and the number of values for each condition factor is regarded as the level number of that factor. Firstly, based on the specifications of the tested software, we identify the operational elements and external factors that impact its functionality, treating them as factors. Secondly, we determine different values for these factors and establish their level numbers before selecting an appropriate orthogonal table. Finally, we use this table to combine the states of each factor in order to construct an effective test input data set. The specific steps are outlined below.

(1) Clarify what factors (variables) are available.

(2) Which levels (values of variables) are available for each factor.

(3) Select an appropriate orthogonal table.

(4) Map the values of the variables to the table.

(5) Use the combination of levels of each factor in each row as a test data.

(6) Some other test data may be added.

Many orthogonal tables have been publicly released, and can be obtained from the Internet, mathematical statistics books, related software, and other sources for standardized orthogonal tables. In selecting a suitable orthogonal table, the number of factors (variables), the number of factor levels (values of variables), and the number of rows of the orthogonal table need to be considered. In the case where there are more than one orthogonal table that meets the need, the one with the least number of rows should be taken. If the number of factors (variables) and levels (values of variables) match, then it is sufficient to apply the orthogonal table that meets the need directly. If the number of factors and levels do not match the orthogonal table, the following principles can be followed.

(1) The number of columns in the orthogonal table cannot be smaller than the number of factors.

(2) The number of levels of the orthogonal table cannot be smaller than the maximum number of levels of the factors.

(3) Orthogonal tables have as few rows as possible.

The above principles can be summarized as minimum inclusion. It is to find the orthogonal table that is greater than or equal to the desired number of factors and number of factor levels with the minimum number of rows.

5.8.4　Example of application of the orthogonal experiment method

正交实验法应用
示例

Λ system has 5 independent parameter configuration variables (A, B, C, D,

E), both variables A and B have 2 values (A1, A2) and (B1, B2). Variables C and D have 3 possible values (C1, C2, C3 and D1, D2, D3) Variable E has 6 possible values (E1, E2, E3, E4, E5, E6). It is now required to test the execution of the system under different parameter configurations.

If all possible parameter configurations combination are tested, 216 ($2 \times 2 \times 3 \times 3 \times 6$) tests are required. To reasonably reduce the number of tests, orthogonal experiment method can be used. When selecting the orthogonal table, the following conditions are required to be met.

(1) Number of factors $\geqslant 5$.

(2) The number of levels should meet the following requirements.

① The number of levels with 2 factors $\geqslant 2$.

② The number of levels with 2 factors $\geqslant 3$.

③ The number of levels with 1 factor $\geqslant 6$.

There are 2 orthogonal tables that satisfy the conditions above: L_{49} (7^8), L_{18} ($3^6 6^1$), according to the principle of fewer rows, L_{18} ($3^6 6^1$) should be selected. After selecting L_{18} ($3^6 6^1$), since there are only 5 actual variables and this orthogonal table has 7 factor columns, the redundant columns in the orthogonal table should be deleted, as shown in Figure 5-29, and note that the seventh column with a level of 6 cannot be deleted.

experiment number \ factor	1	2	3	4	5	6	7
1	0	0	0	0	0	0	0
2	0	0	1	1	2	2	1
3	0	1	0	2	2	1	2
4	0	1	2	0	1	2	3
5	0	2	1	2	1	0	4
6	0	2	2	1	0	1	5
7	1	0	1	2	1	2	5
8	1	0	2	0	2	1	4
9	1	1	1	1	1	1	0
10	1	1	0	2	0	0	1
11	1	2	1	1	2	0	3
12	1	2	0	0	0	2	2
13	2	0	1	2	0	1	3
14	2	0	2	1	1	0	2
15	2	1	0	1	0	2	4
16	2	1	1	0	2	0	5
17	2	2	0	0	1	1	1
18	2	2	2	2	2	2	0

experiment number \ factor	1	2	3	4			7
1	0	0	0	0			0
2	0	0	1	1			1
3	0	1	0	2			2
4	0	1	2	0			3
5	0	2	1	2			4
6	0	2	2	1			5
7	1	0	1	2			5
8	1	0	2	0			4
9	1	1	1	1			0
10	1	1	0	2			1
11	1	2	1	1			3
12	1	2	0	0			2
13	2	0	1	2			3
14	2	0	2	1			2
15	2	1	0	1			4
16	2	1	1	0			5
17	2	2	0	0			1
18	2	2	2	2			0

Figure 5-29　Delete redundant factor columns

The variable mapping is performed as follows.

A. 0→A1, 1→A2.

B: 0→B1, 1→B2.

C: 0→C1, 1→C2, 5→C3.

D: 0→D1, 1→D2, 5→D3.

E: 0→E1, 1→E2, 5→E3, 3→E4, 4→E5, 5→E6.

The number of values of some variables is less than the number of levels on some factor, and it is necessary to replace the values that are not present with the values that are present evenly, as shown in Figure 5-30.

Figure 5-30   Replace the values that are not present

Exercise 5

I. Single choice questions.

1. Speculate the possible errors based on experience or intuition, list the possible errors in the program or the special circumstances prone to errors, so as to select the test case, such a test method is called (   ).

    A. equivalence class division   B. boundary value analysis

    C. error guess method   D. logical coverage test

2. Black-box test techniques do not include (   ).

    A. equivalence class division   B. boundary value analysis

    C. erroneous guessing   D. logical coverage

3. The most widely used use case design technique for black-box test is (   ).

    A. equivalence class division   B. boundary value analysis

    C. error guess   D. logical coverage

4. In a university student registration management information system, assuming that the input range of student age is 16-40, the following division is correct according to the equivalence class division technique in black-box test. (   )

    A. can be divided into 2 valid equivalence classes and 2 invalid equivalence classes

    B. can be divided into 1 valid equivalence class and 2 invalid equivalence classes

    C. can be divided into 2 valid equivalence classes and 1 invalid equivalence class

    D. can be divided into 1 valid equivalence class and 1 invalid equivalence class

5. A set of test cases that causes each branch of the program under test to be executed at least once satisfies the coverage criterion is (   ).

    A. statement coverage   B. decision coverage

    C. condition coverage   D. path coverage

6. When determining a black-box test strategy, the preferred method is (   ).

    A. boundary value analysis method   B. equivalence class division

    C. error inference method   D. decision table method

7. The (   ) method designs test cases based on the dependency of outputs on inputs.

    A.path test   B. equivalence classes

    C. cause-effect diagram   D. generalization test

8. For the parameter configuration class of software, use (   ) to select fewer combinations to achieve the best results.

    A. equivalence class division   B. cause-effect diagram method

    C. orthogonal test method   D. scenario method

9. For a clear business flow system can be utilized (   ) throughout the test case design process and use a variety of test methods in the use case.

    A. equivalence class division   B. cause-effect diagram method

    C. orthogonal test method   D. scenario method

10. With the boundary value analysis method, assuming $1 < X < 100$, then the integer X in the test should take the boundary value does not include (   ).

    A. $X=1$, $X=100$ B. $X=0$, $X=101$ C. $X=2$, $X=99$ D. $X=3$, $X=98$

II. Fill in the blanks.

1. There are two different cases of equivalence class division: _____ and

_____.

2. If there are multiple input conditions and there are associations between the conditions, it is not enough to cover all the equivalence classes, but you also need to consider the _____ between the equivalence classes.

3. The total number of equivalence classes for each variable under test is equal to its _____ plus _____ .

III. True or false questions.

1. A test case can cover multiple valid and invalid equivalence classes. (   )

2. The quality of test cases obtained from different equivalence class divisions varies.(   )

3. In robust equivalence class test, the number of test cases is: the sum of the total number of equivalence classes for each variable under test. (   )

IV. Comprehensive questions.

1. A certain message encryption code consists of three parts, and the names and contents of the three parts are shown as follows.

Encryption type code : blank or three digits;

Prefix code : a three-digit number that does not start with "0" or "1";

Suffix code : four digits.

Assuming that the program under test can accept all information encryption codes that meet the above requirements and reject all information encryption codes that do not meet the requirements, try to analyze all its equivalence classes and design test cases.

2. A "banking website system" login interface is shown in the Figure 5-31, try to use the error guess method, cite 10 kinds of common problems or errors, and design 10 test cases.

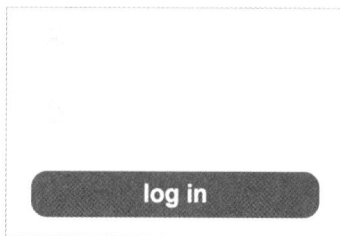

Figure 5-31   Banking website system

3. There is an online shopping website system, the main functions of which include logging in, shopping for products, and paying online to complete the shopping, etc. Users may encounter problems when using these functions. Users may experience various situations when using these functions, such as account not existing, wrong password, insufficient account balance, etc. Let's say there is only one account abc in the system; the password is 123; the account balance is 200; there

are only goods A, both priced at 50 yuan, with an inventory of 15, and goods B, priced at 50 yuan, with an inventory of zero.

Try the scenario approach:

(1) Analyze and draw an event flow diagram, labeling the basic and alternative flows.

(2) Analyze to generate test scenarios.

(3) Design appropriate test cases for each scenario.

4. A software requirement specification contains the following requirement: the first column of characters must be either A or B, and the second column of characters must be a number, in which case file changes are made. However, if the first column character is incorrect, the message L is output; if the second column character is not a number, the message M is given. Please analyze this using a causal diagram and draw a causal diagram corresponding to this software requirement specification statement.

5. There is the following program with variable $x \in (100, 200]$. Please analyze its equivalence class and design test cases to implement the equivalence class division test.(Note: Test cases should include test inputs and expected outputs.)

```
short Func(int x)
{ short ix.
    ix = 0; if((x <= 100))
    if((x <= 100)||(x > 200))
{ printf("Variable x input data out of range! \n");
  return ix; }
          else if(x < 150)
              { ix= 3; }
              else if (x == 200)
              { ix = 4; }
              else ix = 5; }
  return ix; }
```

6. A program functions to output the date of tomorrow for a certain input date, e.g., input February 2, 2020, the output of the program will be February 3, 2020. The program has three input variables year, month, and day that represent the year, month, and day of the input date.

(1) Please classify the input variables year, month, and day as valid equivalence classes according to the program specifications.

(2) Analyze the specification of the program and give all the possible actions that the program can take in the context of the above division of equivalence classes.

(3) Based on (1) and (2), draw a simplified decision table and design test cases for each rule.

7. An online shopping mall, gives different discounts to its members based on their points as shown in Table 5-13. Let the points take integer values, design the test data by using the boundary value method.

Table 5-13   Online shopping mall's different discounts

Level	Points amount	Discount/%
0	0	None
1	1-99	5
2	100-999	10
3	1,000-4,999	15
4	5,000-9,999	20
5	10,000-99,999	25
6	100,000 and above	30

Chapter 6   Black-box test practices

The test cases generated by black-box test design must be executed on the software under test. The actual execution results should be compared with the expected correct results to determine whether the test passes. In the past, these tasks were performed manually, but they are now typically carried out using automated black-box test tools.

6.1   Fundamentals of automated black-box test

自动化黑盒测
试的基本原理

If we want to automate some process, we can usually do so by writing code, such as the following Java code for updating data.

```
...
   connect=DriverManager.getConnection(sConnStr,"sa","123");
stmt=connect.createStatement();
   stmt.executeUpdate(sql);
   stmt.close();
   connect.close();
   ...
```

The code above, can complete the database connection, SQL execution, and connection closing. If you execute this code a few times, you can repeat this process automatically. Similarly, we can automate the execution of black-box test by program code, which is called test scripting. For example, here is a test script for automating the execution of the software under test. There are comments for each statement line to make it easy to be understood.

```
   startApp("ClassicsJavaA");     // Start the app ClassicsJavaA
   tree2().click(atPath("Composers->Bach->Violin Concertos"));
   // Select Composers, Bach, Violin Concertos in the displayed
directory tree
   ...
   placeAnOrder().inputKeys("{Num3}{Num4}{Num1}{Num2}{Num3}{Num4
```

```
}");
    // Enter the number "341234"
    Ok ().click(); // click the "OK" button classicsJava(ANY,
MAY_EXIT).close();
    // Close application ClassicsJavaA
```

This code is executed repeatedly to automatically repeat the test process.

A test script is a set of instructions that can be executed in a test tool, representing a form of computer program. It enables the automatic control of the test process through its execution. Test scripts can be written directly using scripting languages, similar to writing other high-level language programs; however, this requires the writer to have familiarity with the specific scripting language. Alternatively, software testers who are not proficient in scripting languages can easily obtain test scripts through recording technology.

6.1.1   Script record

脚本录制

Script record refers to the manual execution of the test process in a test tool software that supports script record. During this process, the test tool software converts each step of the test operation into test script code and records it. Ultimately, a test script is obtained that can automatically complete the entire test process, as illustrated in Figure 6-1.

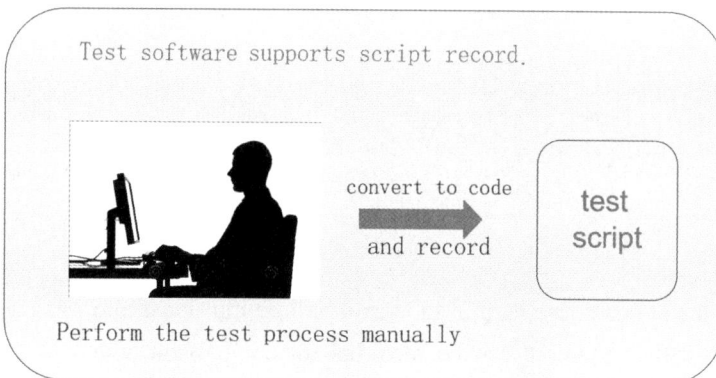

Figure 6-1   Test script record

Recording a test script can significantly reduce scripting efforts by capturing every step of the user's operation, including the location or object of the action. The location refers to the pixel coordinates of the user interface, while objects of operation can include windows, buttons, scroll bars, and more. Actions such as inputting data, clicking on elements, triggering events, changing states or properties are all recorded and converted into a set of instructions or scripts using a scripting language.

Broadly speaking, scripts can be divided into the following types.

(1) Linear scripts: Scripts that are executed sequentially. This is usually done by recording scripts directly from the manual execution of the test process.

(2) Structured script: Similar to high-level language program, it is a script with various logical structures (sequence, branch, loop), and can have function calling.

(3) Data-driven scripts, keyword-driven scripts, shared scripts, and so on.

6.1.2　Script replay 脚本回放

Once the script has been recorded, you can repeat the test by simply executing the script, a process known as replay (as shown in Figure 6-2). In essence, replay automatically repeats the test process by executing the test script.

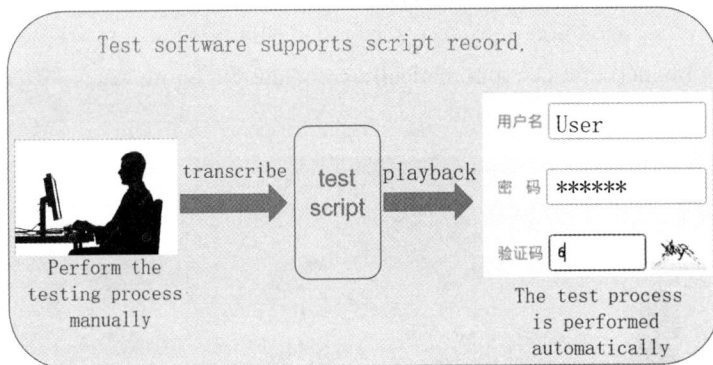

Figure 6-2　The recording and replay process

During script replay, the actions described in the script will be translated into on-screen actions, and the software output can be recorded for comparison with expected results to determine whether the test passes or fails. Through script replay, automated execution of the test process can significantly reduce workload. Additionally, it serves as an effective regression test during iterative development processes.

6.2　Techniques of automated black-box test

自动化黑盒测
试技术

In the process of automated black-box test, various related technologies are utilized, including script optimization, data verification points, data-driven techniques, and virtual users. Only by understanding and mastering these technologies can we achieve automated black-box test.

6.2.1　Script optimization

脚本优化

The scripts generated from recording can be modified and optimized. During the recording process, certain operations that are not relevant to the test, such as mouse sliding, may also be recorded in the test script. These contents can be removed in order to enhance the efficiency of the test.

For instance, in a rational function tester script, the following lines of code are irrelevant to the test and should be eliminated.

```
...
MemberLogon (). DragToScreenPoint (atPoint (209, 9),
toScreenPoint (209, 10));
    // useless window drag
    ClassicsCD (). DoubleClick (atPoint (533368)); // useless mouse
//double click
    ClassicsCD (). Click (atPoint (515320)); // useless mouse click
...
```

We can incorporate logical structures such as branches, loops, and function calls into the test script, similar to structured programming, in order to enhance the functionality of the test scripts.

6.2.2　Data validation points

数据验证点

With the aid of inserting data validation points in the script, data verification can be performed during script playback to determine whether the intermediate or final test results are correct. For example, the following rational function tester script includes data validation points.

```
public class OrderBachViolin extends OrderBachViolinHelper
```

```
{ public void testMain(Object[] args)
  {   startApp("ClassicsJavaA");
      tree2().click(atPath("Composers->Bach->Location(PLUS_
MINUS)"));
      tree2().click(atPath("Composers->Bach->Violin Concertos"));
      placeOrder().click();

      ok().drag();
      QuantityText (). Click (atPoint (35 (9));
      placeAnOrder().inputKeys("{Num1}{Num0}");
      ...
      // the next row insert data verification point,
      // test whether is equal to the total amount
      // of the tested software to calculate the expected value:
      _15090().performTest(OrderTotalAmountVP());
      ...
      placeOrder2().click();
      Determine ().click();
      classicsJava(ANY,MAY_EXIT).close();
  } }
```

Data verification points can not only determine whether the test process or result is correct, but also realize the synchronization between the script code execution and the interface display. For example, the test flow is: after the execution of the former screen, the latter screen is popped up, and then the "ok" button is clicked on the latter screen. However, when the script code executes to click the "ok" button in the latter interface, the "ok" button even the latter interface may has not been displayed yet. In this case, before the code line of clicking the "ok" button, a data verification point should be inserted to check whether the "ok" button in the latter interface has been displayed. The test script is as follows.

```
...
tree2().click(atPath("Composers->Bach->Location(PLUS_MINUS)"));
tree2().click(atPath("Composers->Bach->Violin Concertos"));
placeOrder().click();
// Insert data validation points to verify that the "ok" button is
// displayed on the next action Screen
placeOrder2().performTest(okButtonPropertiesVP()); ok().click();
...
```

6.2.3　Data-driven

数据驱动

It is not meaningful to repeatedly execute the test script without any changes. By configuring the data driver for the test script, different test data can be input each time the test script is executed, enabling the automatic execution of a large number of tests. The data-driven test script involves setting the input data as variables in the script and configuring a variable value table, as illustrated in Figure 6-3. Once the data-driven approach is established, each row in the variable value table can be sequentially used as input data during the execution of the test script.

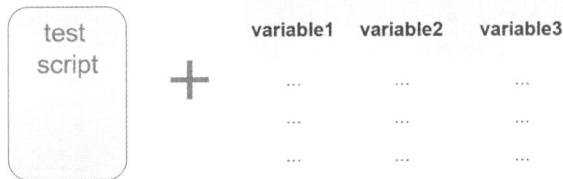

Figure 6-3　Data-driven

With a data-driven test script, although the test process is the same, the test input data is not the same, and each test can achieve different test purposes.

6.2.4　Virtual user technology

虚拟用户技术

In the context of performance test, it is often imperative to assess whether the software under scrutiny can meet the actual requirements when accessed concurrently by multiple users. In such scenarios, tools are commonly employed to simulate the behavior of multiple users. LoadRunner, a performance test tool, refers to this simulation as a virtual user and utilizes virtual user technology for this purpose.

LoadRunner offers several virtual user techniques that enable the generation of server load across different client/server architectures. Each technique is tailored to a specific architecture and results in a distinct type of virtual user. For instance, the Web virtual user can be utilized to replicate a user's actions within a web browser. These various virtual user technologies can be used independently or in combination to create an effective load test scenario.

Exercise 6

Ⅰ. Single choice questions.

1. (　　) can make the software testers who are not familiar with the script language can also easily get the test script.

    A. Recording technology   B. Playback technology

    C. Data driven technology   D. Data verification point technology

2. In automated black-box test tools, after the script is recorded, you can replay the test process by simply executing the script, which is called (　　).

    A. replaying   B. replaying   C. copying   D.replaying record

3. (　　) is not an advantage of automated software test.

    A. High speed and efficiency

    B. High accuracy and precision

    C. Achieving some manual difficult to complete the test

    D. Fully and thoroughly test the software

Ⅱ. Fill in the blanks.

1. The data drive of the test script in the test tool is to set _____ as a variable in the script.

2. In automated black-box test, the technique used to check the results is _____.

3. In automated test, the operation process and operation behavior of the computer are generally used to realize automated black-box test through the special program _____designed.

4. Generally speaking, automated black-box test tools can only improve the efficiency of test _____, but not necessarily improve the effectiveness of test.

Ⅲ. True or false questions.

1. In order to test a banking APP, we should collect a large number of users' bank account numbers, ID numbers, phone numbers and other information for test work. (　　)

2. When using performance test tools, a large number of testers are required to simulate the actions of users. (　　)

3. Automated tests can be run regardless of commuting hours, or even 24 hours a day. (　　)

4. We can use software test tools for government websites, data centers, etc., in order to exercise and improve our software test capabilities. (　　)

Ⅳ. Comprehensive questions.

Choose an automated black-box test tool to complete the script recording and playback process for a test task.

Chapter 7 Web test

7.1 Web automated test introduction

Let's explore Web automated test using open source technologies. In order to realize Web automated test, it is necessary to build a more complex test environment. In order to facilitate readers to download relevant software and practice the technology, we combine the National College Software Test Competition to explain.

The competition website can be found at http://www.mooctest.org/, while technical support for the competition is available at http://www.mooctest.net/.

7.1.1 Install Java, Eclipse, and Selenium

1. Install Java

The Java environment is divided into JDK and JRE. JDK is Java development kit, which is an SDK for developers to use and provides a Java development environment and running environment. JRE is Java runtime environment, refers to the running environment of Java. It is oriented to Java program users. Java can be downloaded from the website as follow.

```
http://www.oracle.com/technetwork/Java/javase/downloads/
index.html
```

After downloading and installing, make sure to configure the environment variables as follows.
- JAVA_HOME: C:\Program Files\Java\jdk1.8.0_191\.
- CLASS_PATH:; %JAVA_HOME%\lib\dt.jar; %JAVA_HOME%\lib\tools.jar.
- PATH: %JAVA_HOME%\bin; %JAVA_HOME%\jre\bin.

After the environment variables are configured, press Windows+R key, input "cmd" with keyboard and press enter, open the command window. Please input "java" and press Enter key, then input "javac" and press Enter key, you can verify whether Java is installed successfully. Note that in order to prevent the test script can not be executed correctly due to software version differences, it is recommended to use JDK1.8 version.

2. Install Eclipse

安装 Eclipse

You can directly download Eclipse with the Mooctest plug-in from the technical support website of the National College Students Software Test Competition "http://www.mooctest.net/", or you can choose to install Eclipse by yourself.

3. Install Selenium

安装 Selenium

You can obtain the selenium-standalone.jar file from the technical support website of the National College Software Test competition and then proceed to copy it to C:\mooctest. Alternatively, you may create a new folder if one does not already exist.

4. Install Google Chrome

安装 Google Chrome

Download and install Google Chrome, and configure the environment variables. In the system variables, create a new variable named webdriver. chrome.bin with the value of the file path of "chrome.exe".

5. Install the appropriate version of chromedriver

安装对应版本的 chromedriver

Download the chromedriver_win32.zip file for your Google Chrome version, unzip it, and add the path to the environment variable (add, not new). Create a new variable named webdriver.chrome.driver with the path of the exe file you extracted from chromedriver_win32.zip. Note that the version of chromedriver must match the version of the Google Chrome browser, or you won't be able to open Google Chrome with the test script.

7.1.2   Loading the Selenium jar in Eclipse

在 Eclipse 中加载 Selenium jar 包

After downloading selenium-standalone.jar, unpack the Selenium package, as shown in Figure 7-1.

Open Eclipse and click File→New→Project, and click "Next" button to create a Java project in Eclipse, as shown in Figure 7-2.

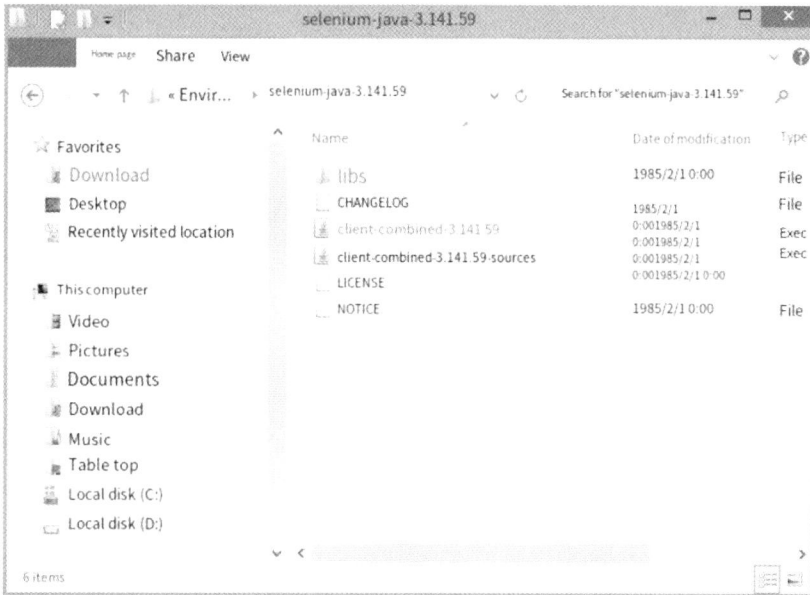

Figure 7-1　Decompression Selenium package

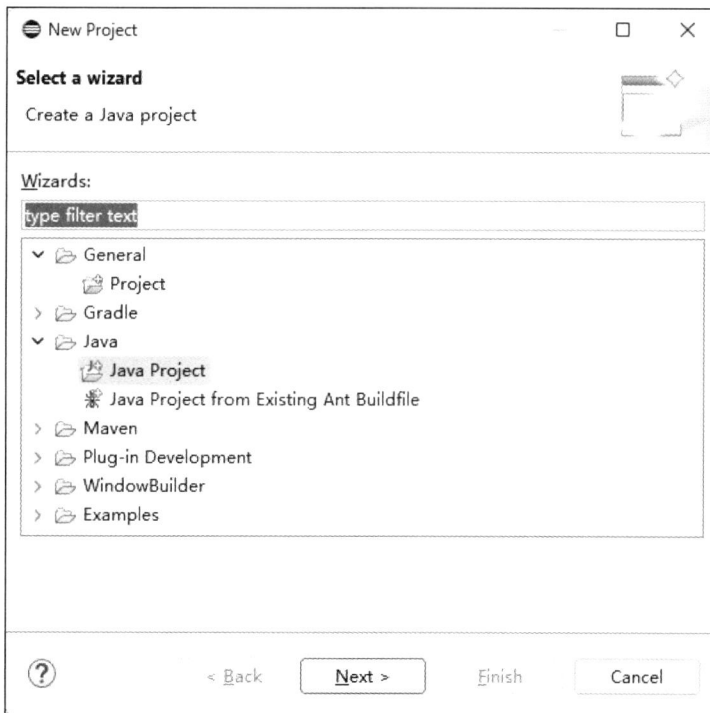

Figures 7-2　Click "Next" button

Input project name SeleniumTest1, and click "Finish" button, as shown in Figure 7-3.

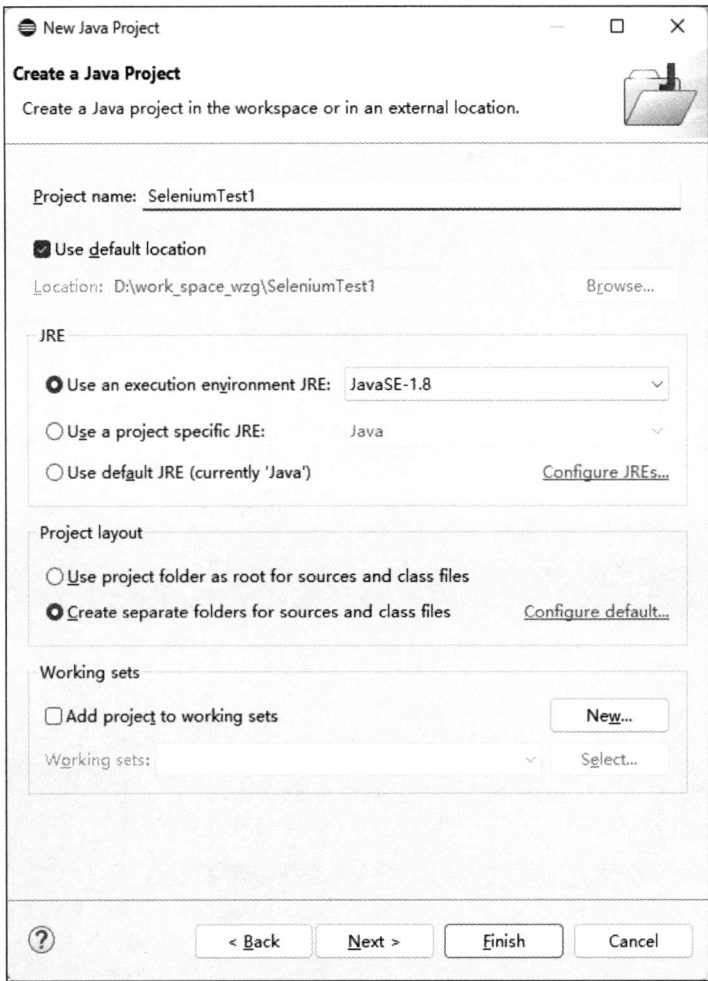

Figure 7-3　Input project name SeleniumTest1, and click "Finish" button

Create a package (The concept of a package can be understood as a collection of programs) and a class on SeleniumTest1 with a package name of "com.SeleniumLib.jase" and a Class name of "T1.java"(Figure 7-4).

Let's import the Selenium packages we need, which are the Selenium jars we downloaded and unzipped earlier, and add all the jar files under it. You can use Ctrl+A to select all files in one directory at a time, including all jars in the libs directory.

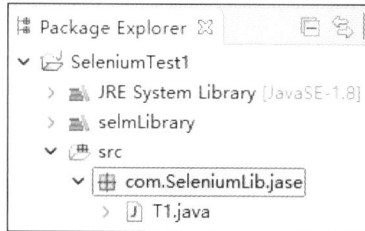

Figure 7-4   Creating a package and a class

To do this, right click SeleniumTest1→Build Path→Add Libraries, and click "Next", as shown in Figure 7-5.

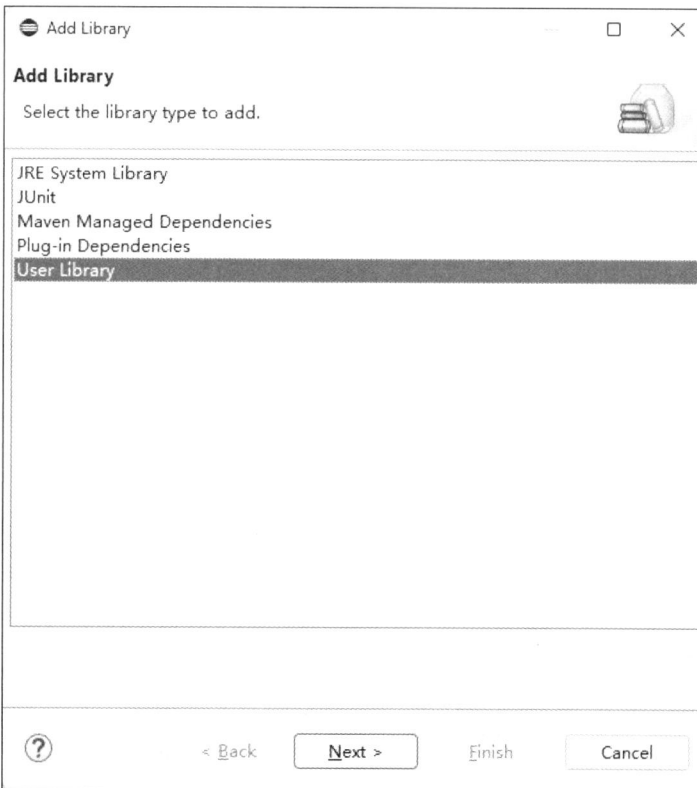

Figure 7-5   Add library

Click "Use Libraries..."→"New...", input user library name "SeleniumLib", click "OK".

Click "Add External JARs...", select the file "selenium-standalone.jar", and click "Open".

Click "Apply and Close", as shown in Figures 7-6.

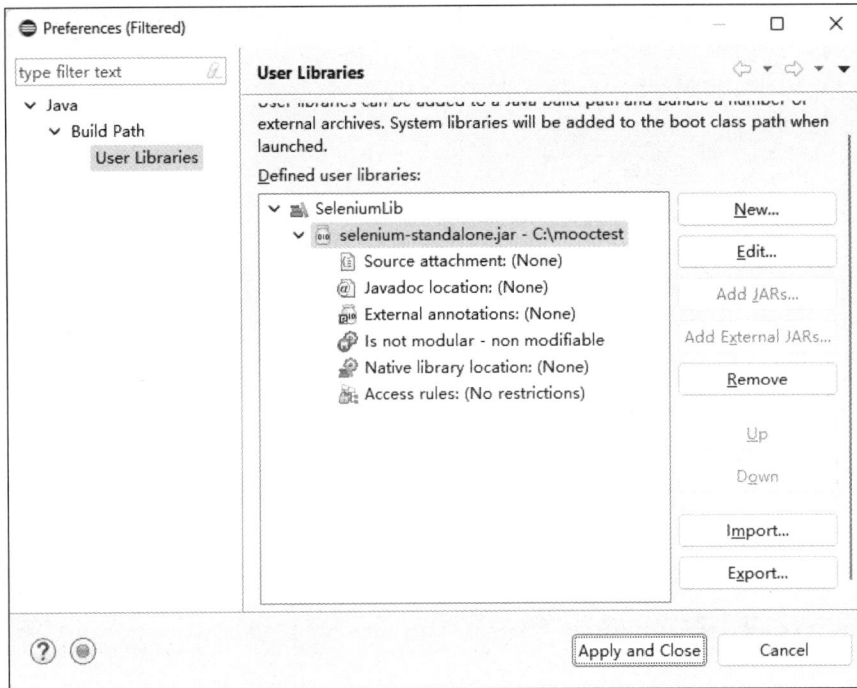

Figure 7-6   Click "Apply and Close"

At last click "Finish", and then the Selenium jar has been loaded in Eclipse now.

7.2   Web test practices

Web 测试实践

7.2.1   Basic test operations

基本测试操作

1. Web page jump

网页跳转

With driver.get (" ******"), we can navigate to the specified Web page. The following line of code is used to jump to Baidu home page.

```
driver.get("https://www.baidu.com/");
```

2. Get the ID of the object under test

获取被测对象 ID

Move the mouse cursor over the component of the web page, and right-click it, a right-click menu will appear. For example, on the Baidu homepage, point the mouse at the "Baidu baid" button, and the interface shown in Figure 7-7 will appear.

Figure 7-7　Opens the right-click menu of a Web page

Then click "Inspect" and a code window appears on the right. You should see the line of code for the "Baidu baid" button, along with its ID (Figure 7-8).

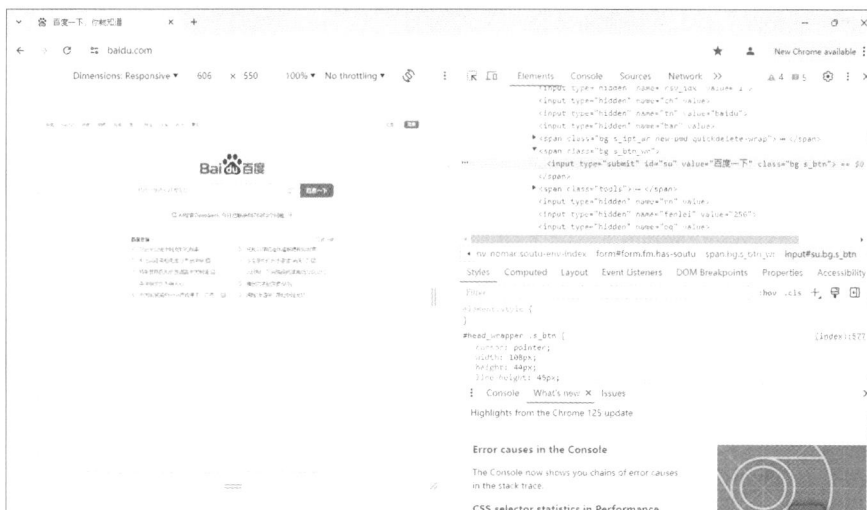

Figure 7-8　The code window showing the object line along with its ID

Copy the ID and paste it into the test script as follows.

```
driver.findElement(By.id("su")).click();
```

3. Get the XPath of the object under test

Move the mouse cursor over the component of the Web page, right-click, and a right-click menu will appear, then click "Inspect", then in the code window on the right, click the three small dots to the left of the blue check bar, and then click "Copy", "Copy XPath" in the menu that appears, as shown in Figure 7-9.

Note that the XPath information of the web component is copied to the clipboard after clicking, and there is no direct feedback on the result.

获取被测对象 XPath

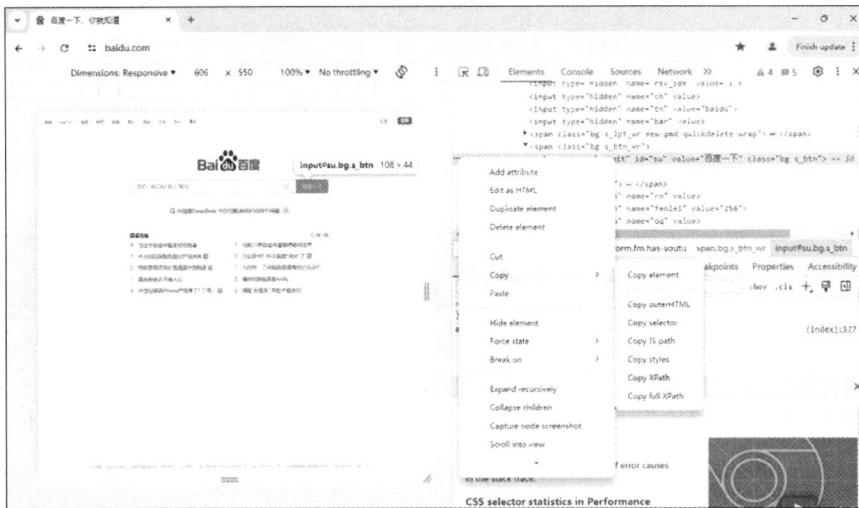

Figure 7-9　Copy XPath

With XPath copy, the web component's XPath information is copied to the clipboard, and then pasted in the test script to obtain the component's XPath.

```
driver.findElement(By.xpath("//*[@id=\"su\"]")).click();
```

4. Simulate mouse click operation

You can locate the element by id and click it as follows.

```
driver.findElement(By.id("su")).click();
```

You can locate the element by XPath and click it as follows.

```
driver.findElement(By.xpath("//*[@id=\"su\"]")).click();
```

实现模拟鼠标单击操作

5. Simulate keyboard input

模拟键盘输入操作

To simulate keyboard input, use sendKeys(). Here are script examples.

```
// This code is used to locate the element by id and perform
// input on it.
driver.findElement(By.id("#loginform>div:nth-child(1)>input"))
.sendKeys("admin");
// This code is used to locate the element by xpath and perform
// input on it.
driver.findElement(By.xpath("//*[@id=\"loginform\"]/div[2]/
input")).sendKeys("123456");
```

6. Wait

等待

You can wait with Thread.sleep().

```
//put the thread to sleep to finish loading the page and prevent
// missing elements.
Thread.sleep(2000);  // 1000=1 秒
```

7. Maximize the web page window

最大化网页窗口

You can maximize the web page window by using maximize().

```
// Maximize the web page window
driver.manage().window().maximize();
```

8. Close the browser window

关闭浏览器窗口

You can close the browser window with driver.quit().

```
Public static void main(String[] args) {
    // Run main function to test your script.
  WebDriver driver = new ChromeDriver();
    try { test(driver); }
        catch(Exception e) { e.printStackTrace(); }
    // close the browser window when the test is complete.
    finally { driver.quit(); }
}
```

7.2.2   Getting started practical case

入门实践案例

After introducing the test environment construction, tool installation and basic test operations, let's look at one of the simplest introductory practice cases. The test operation completed by this case is to open the Baidu homepage with the test script, enter the search keyword "software test", click "Baidu once" to start searching, wait for 5s to complete the search and view the results. At last, close the browser.

1. Create a new class

In the Java project SeleniumTest1 that you created earlier, create a new class called Web_test1 (Figure 7-10).

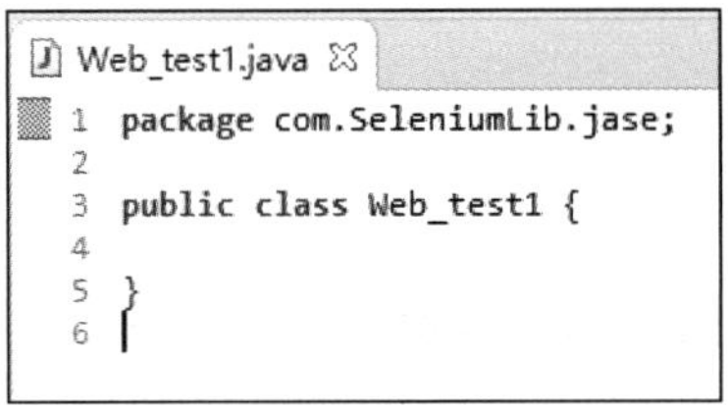

Figure 7-10   Create a new class called Web_test1.java

You can also directly download the configured Web test project from the technical support website of the National College Software Test Competition http://www.mooctest.net/.

2. Import the package files required for web test

Import the files needed for web test in Web_test1.java, as shown in Figure 7-11.

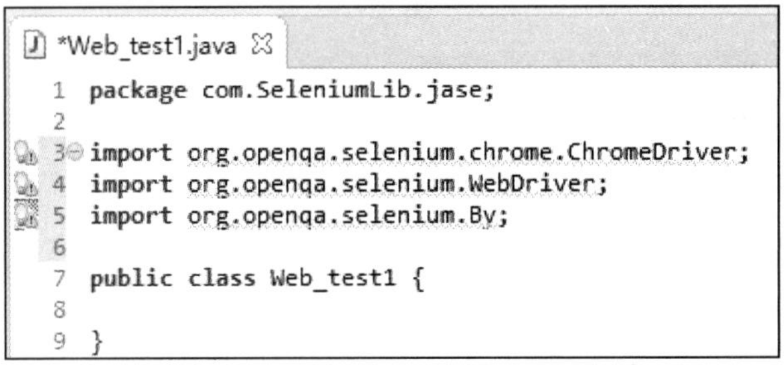

Figure 7-11   Import the files needed for web test

3. Create test method

Create the test method under class Web_test1 as follows, with comments given in the code.

```
public static void test(WebDriver driver) {
try {
// maximize the web page
driver.manage().window().maximize();
```

```
// navigate to the page
driver.get("https://www.baidu.com/");
// wait for the web page to load
Thread.sleep(1000);
// Enter a search term.
driver.findElement(By.id("kw")).sendKeys(" Software test ");
// Click the "Baidu Once" button to start the search.
driver.findElement(By.id("su")).click();

// wait in order to finish the search and see the search results
Thread.sleep(10000);

} catch (Exception e) {
e.printStackTrace();
}
}
```

4. Create the main function

Create the main function under class Web_test1 as follows, with comments given in the code.

```
public static void main(String[] args)
 {
// Define WebDriver named driver
WebDriver driver = new ChromeDriver();

// Test driver, open browser
try { test(driver); }
catch(Exception e) { e.printStackTrace(); }

// Close the browser and exit the test
finally { driver.quit(); }
}
```

5. Execute the test script and get the result

When Executing the test script will start the Chrome browser, open the Baidu homepage, enter the keyword "software test" in the search input field, and start searching, as shown in Figure 7-12.

Figure 7-12   Search results after the test script is executed

When the last wait time set in the test script is over, the browser will be closed and the test will exit.

7.2.3   Advanced practice example

Let's look at a complicated practice example. This case is a practice question of the National College Software Test Competition.

1. Test requirements

Write a script to implement the Web functional test, test URL is https://www.suning.com/.

When executing the test script, it is important to consider that the response time of the web page may vary due to factors such as network speed and server load. Sufficient waiting time should be incorporated into the test script to ensure smooth loading of the Web page. During the process of opening the page, there may be instances such as pop-up windows or verification prompts. In such cases, manual intervention may be necessary to close these prompts or re-run the code in order to ensure that subsequent processes operate normally. If there are scenarios requiring manual verification, code can be written to facilitate this process, or manual test can be conducted in order to pass verification.

The specific test requirements are as follows.

(1) Open the home page of Suning Tesco, maximize the window, and select the city as Zhenjiang.

(2) Select [Classification].

(3) Select [Suning Appliance].

(4) Choose a hanging air conditioner.

(5) Select [Midea].

(6) Select number of pieces of goods [3 pieces].

(7) Select price [0-7000].

(8) Select the first information picture to jump.

(9) Enter [Gree Air Conditioning] in the input box and click [Search] to search.

(10) Select [Household Air Conditioning] [2pi] [New Level 3] in turn.

(11) Select [5000-6000] [wall-mounted air conditioning] [Master bedroom].

(12) Select the third data picture of the search results.

(13) Select [Add to Cart].

(14) Select [Go to Cart to Settle].

(15) Select [Delete].

(16) Select [OK].

2. Description of the test script writing process

For such test requirements, the process of the test script writing is the same as the "7.2.2" , but it is more complex and the workload is larger. Limited by space, it is not described in detail here.

3. Test script

According to the test requirements, write the test script as follows.

```
import org.openqa.selenium.chrome.ChromeDriver;
Import org.openqa.selenium.support.ui.ExpectedConditions;
import org.openqa.selenium.support.ui.WebDriverWait;
import com.gargoylesoftware.htmlunit.javascript.host.Set;
import com.sun.org.apache.bcel.internal.generic.Select;
import com.sun.xml.internal.bind.v2.schemagen.xmlschema.List;
import org.openqa.selenium.WebDriver;
import org.openqa.selenium.WebElement; import java.util.
```

```
ArrayList;
    import org.openqa.selenium.By;


    public class Example {
    // Mooctest Selenium Example
    // <! > Check if selenium-standalone.jar is added to build path.
    public static void test(WebDriver driver) {
    // TODO Test script
    // eg:driver.get("https://www.baidu.com/")
    // eg:driver.findElement(By.id("wd"));
    try { driver.get("https://www.suning.com/"); driver.manage().
window().maximize(); Thread.sleep(500);
    driver.findElement(By.id("citybName")).click();
    Thread.sleep(1500);
    driver.findElement(By.name("item_none_dizhi_02")).click();
     Thread.sleep(1500);
    driver.findElement(By.linkText(" Jiangsu ")).click();
    // Jiangsu
    Thread.sleep(3000);
    driver.findElement(By.name("item_none_dizhi_03")).click();
    Thread.sleep(1500);
    driver.findElement(By.linkText(" Zhenjiang ")).click();
    // Zhenjiang
    Thread.sleep(3000);
    driver.findElement(By.className("index-all-hook")).click();
    //classification
    Thread.sleep(1500);
    driver.switchTo().window((String)
    driver.getWindowHandles().toArray()[1]); // jump
    Thread.sleep(1500);
    driver.findElement(By.name("public0_none_dbdh_dh3")).click();
    // Suning appliance
    Thread.sleep(1500);
    driver.switchTo().window((String)
    driver.getWindowHandles().toArray()[2]); // jump
    Thread.sleep(1500);

    driver.findElement(By.name("dianqi_newhome3_40642214888_wor
d02")).click();
    // Hanging air conditioner
    Thread.sleep(1500);
```

```
   driver.switchTo().window((String)
   driver.getWindowHandles().toArray()[3]); // jump
   Thread.sleep(1500);
   driver.findElement(By.cssSelector("# Midea > a > img")).click();
   //first location picture
   Thread.sleep(1500);
   Driver. FindElement (By xpath (" / / * [@ id = \ "3 horse \"]
/ span")). Click (); // three horses
   Thread.sleep(1500);
   driver.findElement(By.id("0-7000")).click();
   Thread.sleep(1500);
   driver.findElement(By.xpath("//*[@id=\"ssdsn_search_pro_bao
guang-1-0-1_1_01:007 1564281_12114867081\"]/i/img")).click();
// first position image
   Thread.sleep(1500);
   driver.switchTo().window((String)
   driver.getWindowHandles().toArray()[4]); // jump
   Thread.sleep(1500);
   driver.findElement(By.id("searchKeywords")).sendKeys(" Gree
Airconditioner ");
   Thread.sleep(1500);
   driver.findElement(By.id("searchEbuySubmit")).click();
   Thread.sleep(1500);
   driver.findElement(By.linkText(" home air conditioner ")).
click();
   Thread.sleep(1500);
   driver.findElement(By.linkText("2 pieces ")).click();
   Thread.sleep(1500);
   driver.findElement(By.linkText(" new level 3 ")).click();
   Thread.sleep(1500);
   driver.findElement(By.id("5000-6000")).click();
   Thread.sleep(1500);
   driver.findElement(By.linkText(" wall air conditioner ")).
click();
   Thread.sleep(1500);
   driver.findElement(By.linkText(" master bedroom ")).click();
   Thread.sleep(1500);

   driver.findElement(By.xpath("//*[@id=\"0071377308-122261917
39\"]/div/div/div[1]
```

```
/div")).click();
// third position image
Thread.sleep(1500);
driver.switchTo().window((String)
driver.getWindowHandles().toArray()[5]); // jump
Thread.sleep(1500);
driver.findElement(By.id("addCart")).click();
 // Add to cart
Thread.sleep(1500);
```

```
    driver.findElement(By.className("go-cart")).click(); // Go to
// cart
```

```
checkout
```

```
Thread.sleep(3000);
driver.findElement(By.linkText(" delete ")).click(); // delete
Thread.sleep(1500);
driver.findElement(By.name("icart1_goods_suredelate")).click();
//delete
```

```
Thread.sleep(1500);
}catch(Exception e) {
e.printStackTrace();
}}
```

```
public static void main(String[] args) {
// Run main function to test your script.
WebDriver driver = new ChromeDriver();
try { test(driver); }
catch(Exception e) { e.printStackTrace(); }
finally { driver.quit(); }}}
```

It is important to note that this test script is actually executed and can realize the corresponding test requirements. However, the tested web page is under constant maintenance, the page may change, so it is not guaranteed that the reader can still execute this script correctly. The test script needs to be continuously maintained and updated with the software under test.

Exercise 7

Ⅰ. Single choice questions.

1. In the Web test script, the code to wait 5s is (   ).

    A. Thread.sleep(5)   B. Thread.sleep(50)   C. Thread.sleep(500)   D.Thread.sleep(5000)

2. The script code for closing the browser window is (   ).

    A. driver.quit( )   B. window.quit( )   C. driver.close( )   D.window.close( )

3. When you test a Web site, one of the functional tests is (   ).

    A. connection speed test   B. link test

    C. concurrency test   D. security test

Ⅱ. Fill in the blanks.

1. The Web test code for jumping to the Sina home page is:

    _____("https://www.sina.com.cn/").

2. Move the mouse cursor to the component of the web page, click the right button, and the right button _____ will appear.

3. To simulate _____ in a Web test script, use sendKeys( ).

Ⅲ. True or false questions.

1. Web test scripts need to be maintained continuously as the web page under test is updated.(   )

2. Test script execution should be fully automated without human interaction. (   )

3. The Google Chrome browser and chromedriver in the Web test environment are separate and do not interfere with each other. (   )

4. After "Copy XPath" is completed, the XPath information of the web component is copied to the clipboard, and there is no direct feedback of the result. (   )

Ⅳ. Comprehensive questions.

1. Write the operations completed by the following test scripts.

```
driver.get("https://music.91q.com/");
Thread.sleep(3000);

driver.findElement(By.id("search_Key_input")).click();
Thread.sleep(1000);

driver.findElement(By.id("search_Key_input")).sendKeys("Jay");
river.findElement(By.id("search_button")).click();
```

2. Write a test script to realize the following test operations.

(1) Open the website at https://music.163.com/.

(2) Click the "Sign In" button on the home page, then choose your login method to log in.

(3) Click the "Leaderboard" button once you're logged in.

(4) Click on "Hot Chart" and select the number one song on the Hot chart.

(5) Click the "Favorites" button to create a new playlist.

(6) Enter the song name as "MyTest" and click "OK" to finish creating and bookmarking the playlist.

附录 A 中英文对照词汇表

英　文	中　文
action stub	动作桩
arc sequence	弧序列
automated Web test	Web 自动化测试
automatic execution	自动化执行
black-box test	黑盒测试
boundary value	边界值
branch coverage	分支覆盖
cause-effect diagram	因果图
compatibility test	兼容性测试
compiler	编译器
condition combination coverage	条件组合覆盖
condition coverage	条件覆盖
condition stub	条件桩
constraints	约束
control flow graph	控制流图
coverage criteria	覆盖标准
data flow analysis	数据流分析
data verification point	数据验证点
database access error	数据库访问错误
data-driven scripts	数据驱动脚本
data-driven test	数据驱动测试
decision coverage	判定覆盖
decision table driven	决策表驱动
dynamic test	动态测试
embedded software	嵌入式软件
encryption type code	加密类型码
equivalence class division	等价类划分
equivalence table	等价表

续表

英　　文	中　　文
error guess	错误猜测
exploratory test	探索式测试
expression analysis	表达式分析
external data access error	外部数据访问错误
function call relationship diagram	函数调用关系图
inclusive	包含
inheritance relationship	继承关系
interface analysis	接口分析
judgment table driven	判定表驱动
keyword-driven scripts	关键字驱动脚本
linear script	线性脚本
logical coverage	逻辑覆盖
loop complexity	环路复杂度
module coupling	模块耦合
module integration	模块集成
module structure diagram	模块结构图
mutation operator	变异算子
mutually exclusive	互斥
nested loop	嵌套循环
node sequence	节点序列
non-functional features	非功能特性
object-oriented	面向对象
orthogonal experimental design	正交实验设计
parameter	参数
polymorphic test	多态测试
prefix code	前缀码
process-oriented	面向过程
program flow chart	程序流程图
program mutation test	程序变异测试
recording technology	录制技术
regression test	回归测试
scenario method	场景法

续表

英　文	中　文
screen resolution	屏幕分辨率
script optimization	脚本优化
script replay	脚本回放
software defect	软件缺陷
software quality measurement	软件质量度量
software specification	软件规格说明
software test	软件测试
statement coverage	语句覆盖
static test	静态测试
suffix code	后缀码
test case	测试用例
test efficiency	测试效率
test requirements analysis	测试需求分析
test script	测试脚本
the first class software test method	第一类软件测试方法
path exhaustion test	路径穷举测试
the second class software test method	第二类软件测试方法
the set of independent path	独立路径集合
unique	唯一
variable	变量
variable cross-reference table	变量交叉引用表
virtual user	虚拟用户
white-box test	白盒测试

参 考 文 献

[1] 阿尼什 毛里西奥.Effective 软件测试[M]. 朱少民，李洁，张元，译. 北京：清华大学出版社，2023.

[2] 朱少民. 软件测试方法和技术[M]. 4 版. 北京：清华大学出版社，2022.

[3] 王蓁蓁. 软件测试：原理、模型、验证与实践[M]. 北京：清华大学出版社，2021.

[4] 郑炜，刘文兴，杨喜兵，等. 软件测试[M]. 慕课版. 北京：人民邮电出版社，2019.

[5] 林若钦. 基于 JUnit 单元测试应用技术[M]. 广州：华南理工大学出版社 2017.

[6] 李炳森. 实用软件测试 [M]. 北京：清华大学出版社，2016.

[7] 宫云战. 软件测试教程[M]. 2 版. 北京：机械工业出版社，2016.

[8] 朱少民. 软件测试[M]. 2 版. 北京：人民邮电出版社，2016.

[9] 周元哲. 软件测试实用教程[M]. 北京：人民邮电出版社，2013.

[10] 李海生，郭锐. 软件测试技术案例教程[M]. 北京：清华大学出版社，2012.

[11] Hetzel W C. The complete guide to software testing[M]. 2nd ed. Wellesley, MA: Qed Information Sciences, 1988.

[12] Brown S, Timoney J, Lysaght T, et al. 软件测试原理与实践[M]. 英文版. 北京：机械工业出版社，2012.

[13] Myers G J. 软件测试的艺术[M]. 3 版. 张晓明，等译. 北京：机械工业出版社，2013.

[14] Vance S. 优质代码：软件测试的原则、实践与模式[M]. 伍斌，译. 北京：人民邮电出版社，2015.

[15] Koskela L. 有效的单元测试 [M]. 申健，译. 北京：机械工业出版社，2014.

[16] Miles G J, Badgett T, Sandler C. 软件测试的艺术[M]. 张晓明，黄琳，译. 3 版. 北京：机械工业出版社，2012.

[17] Tahchiev P, Leme F, Massol V，et al. JUnit 实战[M]. 王魁，译. 2 版.北京：人民邮电出版社，2012.

[18] Mathur A P. 软件测试基础教程[M]. 王峰，郭长国，陈振华，等译. 北京：机械工业出版社，2011.

[19] 华为.华为公司简介[EB/OL].(2022-06-01)[2025-03-01]. https://www.huawei.com/cn/corporate-information.

[20] 曾岳，董如婵.系统测试分析与设计[M]. 北京：高等教育出版社，2018.